FINANCING THE INTERNATIONAL PETROLEUM INDUSTRY

Financing the International Petroleum Industry

Norman A. White

With the collaboration of

Albert W. Angulo	George S. Miller
Brian A. Carlisle	Otto R. Norland
Patrick F. Connolly	Roy T. Pleasance
Peter M. C. Clarke	Radu G. Plessia
Peter J. G. Elwes	Peter J. Wingett
R. Kenneth Merkey	Philip Wood

amacom

A Division of American Management Associations

First Published in 1978 by
Graham & Trotman Ltd
Bond Street House
14 Clifford Street
London W1, United Kingdom

© Graham & Trotman Ltd, 1978
ISBN 0-8144-5572-7
Library of Congress Catalog Card Number 79-51576

Printed litho in Great Britain

HD
9560.4

.F5
1978

Contents

Contents

Contents

Preface

This was a *team project*. It is the result of nearly three years' research and collaboration between the publishers, the contributors and myself as Project Director.

Even though I bear overall responsibility for the book, we all shared in the pleasures inherent in such a project. But as someone wrote many years ago — 'Pleasure is nothing else but the intermission of pain' — this project has been no exception.

The original idea for the book came from the publishers. My initial task was to work out the plan of the book following extensive consultations with senior personnel in the financial community and oil industry executives representing the various facets of that industry. These consultations were not only helpful in the main purpose but also were encouraging in confirming the need for such a publication. They also brought forward suggestions for many of the contributors; the final selection of the project team was made by me and is drawn from the USA and various countries in Europe, and is thereby representative of the scope and practices of the international petroleum industry.

Purpose and target audiences

The purpose of this book is to provide an overview of the main activities involved in financing the international petroleum industry. Government involvement in the industry is considered primarily from the viewpoint of its impact on companies in the industry. No special attention is given to the financing of government owned companies in the industry.

There are many excellent textbooks and articles on particular techniques and facets of finance in general terms and on the politics of oil; this book is not intended to compete with these publications. However, as we are not aware of any work which addresses itself to the financing of the petroleum industry, we are now attempting to cover this subject with this book.

The principal target audiences to whom this book is directed are executives in non-financial positions in the oil industry; personnel in banks and other financial institutions who wish to understand more about the petroleum industry; civil servants and politicians who need to understand the interplay between financing and the activities of the international petroleum industry and, lastly, academics and students at graduate schools of business who through their research and studies are probing the multi-disciplinary nature of financial management in various industries.

Plan of book

The book is divided into five parts with each part planned to achieve a particular purpose. Part A provides an introduction to the industry. Chapter 1 attempts to give the reader a perspective of the nature and structure of the industry. Apart from highlighting the truly international character of the industry and the interrelationship and interdependence of its various component parts, it also deals in some detail with OPEC and the future levels of crude oil prices. Chapter 2 presents a brief history of the financing of the industry as an introduction to the more detailed discussions in parts B and C.

The two chapters in part B are intended to contrast the quite different approach towards financing adopted by the small independent members of the industry (usually confined to one segment of the industry) compared with the larger international integrated companies who encompass operations in many countries as well as most activities of the industry.

In accordance with the title chosen for part C the emphasis is on non-recourse financing (using the cash flow from the project as the major, if not the only, source for repayment of the loan finance). Each of the four chapters considers the risks associated with projects in a particular facet of the industry in relation to the approach to security for loan finance used by the banks and other financial institutions. Additionally each chapter covers specific aspects of project financing of more general application than the title of the chapter may suggest. To avoid repetition the coordination and integration of the various contributions are achieved through the medium of the subject index.

Part D is intended to show the relevance of taxation and insurance as they apply to the petroleum industry. The technical nature of these subjects precludes giving more than an indication of the key areas and where they apply.

The attitude and philosophy of the commercial banks having been dealt with in previous chapters, the three chapters in part E give an introduction to three other sources of finance — export credits, stock markets, and eurocurrency markets. These chapters are primarily intended to acquaint the non-financial members of the target audiences with the nature and scope of these markets.

Whilst the different parts of the book each have a particular objective, it is recognised that the subjects covered are interrelated and interdependent. The subject index is intended to serve as a means of integrating the various subjects covered in the individual chapters. It has also been specially prepared bearing in mind the different interests and approaches which it is expected will be shown by the target audiences for whom the book is intended.

Acknowledgments

The project team is international in composition and represents, collectively, many years of experience in financing and the petroleum industry. Each member of the team writes with authority about his subject, having been selected because of his experience in both financing and the petroleum industry. I am personally most grateful to them all for their valuable contributions.

Though this was a team effort, each of us was primarily responsible for one chapter as indicated by the attributed authorship. The members of the team have expended many hundreds of man-hours of painstaking work on their manuscripts, each of which has been specially prepared. I am personally most grateful to the team for their hard work, patience and cooperation that has made possible the completion of the book. By the nature of such a project, the individual contributions have been subject to many changes in greater or lesser measure to fit the requirements of the book as a whole.

I am particularly indebted to the publishers for their patient counsel and continuous encouragement during the nearly three-year period of the research and preparation of the manuscript for the book.

*   *   *   *   *

Finally, financing techniques and the international petroleum industry are both dynamic, never static. Ideas and conclusions which are entirely logical, based on careful analysis of all data available at time of formulation and writing, can become obsolete before they appear in print. It will be a long time before the 'last word' can be written about financing and/or the international petroleum industry.

6 John Street   Norman A White
London WC1N 2ES   Project Director
England

The Project Team

The Project Director

Norman A White is managing partner of Norman White Associates, a research and consulting organisation established in 1972 to specialise in the international energy situation and in corporate development, project financing and international business. He is also a director and chairman, Executive Committee of Tanks Oil and Gas Ltd, a company holding licences in UK and Netherlands sectors of the North Sea and chairman/board member of a number of other commercial and industrial companies in Europe and North America.

From 1972 to 1976 he was special adviser to the board of Hambros Bank Ltd, London. Prior to 1972 Dr White was mainly with Royal Dutch/Shell Group, initially as a research scientist; subsequent positions embraced planning, engineering, commercial and general management in various countries. Latterly, chief executive, New Enterprises Division (concerned with metals, coal mining, energy and other projects on a global basis) directly accountable to Executive Committee, RD/S Group Board.

Dr White is a visiting member of faculty at the Graduate School of Business, Manchester University and The Administrative Staff College, Henley. He is also a member of the Senate, University of London.

Educated mainly at Universities of London, Philippines and Manchester and at Harvard Business School, USA he holds the degrees of BSc(Eng), MSc, PhD(Econ) and is a chartered engineer in the UK. He is also member of Council, Institute of Petroleum; fellow, Institution of Mechanical Engineers and member, American Society of Petroleum Engineers.

Other Contributing Authors

Albert W Angulo (Liquefied Natural Gas Systems)

Mr Angulo is treasurer of International Systems & Controls Corporation, Houston, Texas, a company engaged in international engineering, manufacturing and financial operations for the development of natural resources including a proprietary process for the liquefaction of natural gas. He is also president, Development Funding Corporation, Houston, Texas, which specialises in the arrangement of project financing. Prior to 1970 when he took his present appointment, he was officer in charge of Latin America for the Fidelity Bank, Philadelphia, USA for four years, and from 1959 through 1966 he was in the International Department of Girard Trust Bank, Philadelphia, USA. Born in Madrid, Spain, Mr Angulo received his university education in the USA at Lehigh University (BS in business administration) and Temple University (MBA in international business). Mr Angulo is a member of the Financial Executives Institute.

Brian A Carlisle CBE (International Integrated Companies)

Mr Carlisle is an oil consultant and deputy chairman, Home Oil UK Ltd, and director, Home Oil Ltd, Calgary. Since 1974 he has also been oil adviser to Lloyds Bank International Ltd. From 1955 until his retirement at the end of 1974 Mr Carlisle was with Royal Dutch/Shell Group mainly engaged on finance activities. After a period in London departments he was appointed finance manager, Burmah-Shell Companies, India in 1960 returning to London in 1964 as deputy group treasurer. From 1970 to 1974 he was regional co-ordinator — Middle East and a director of Shell International Petroleum Co Ltd. Prior to joining Shell Mr Carlisle was for eight years a district commissioner in the Sudan.

Patrick F Connolly (Petroleum Production)

Dr Connolly is vice president in charge of the Energy and Minerals Division, Shawmut Bank of Boston, Boston, USA. Immediately prior to his present appointment he was a vice-president in the petroleum group, Republic National Bank of Dallas, Texas for over four years. After a short period with the Bank of Montreal, Canada Dr Connolly joined Republic National Bank of Dallas in 1972 and was posted to London to start European petroleum operations. From 1973 to 1976 he was seconded to the International Energy Bank Ltd London, as marketing manager. Before entering the field of banking Dr Connolly was with Gulf Oil Corporation in UK, Belgium and Japan.

Born in South Africa he received his university education at University College, Dublin, Ireland and at Fletcher School of Law & Diplomacy, Cambridge, Mass., USA. He holds the degrees of BA(Hons) and PhD.

Peter M C Clarke (Eurocurrency Markets)

Mr Clarke is manager, corporate finance, with United International Bank Ltd, London, a position he has held for some years after joining in 1972. Prior to his present appointment he spent seven years in industry with Associated Electrical Industries (now GEC) and Imperial Chemical Industries working in electrical engineering and olefine production. Educated at the University of Cambridge and at Manchester Business School, he holds the degrees of MA and MBA. He is a chartered engineer in the UK, a member of Institution of Mechanical Engineers and an associate member of the Institution of Electrical Engineers.

Peter J G Elwes (Independent Oil Companies)

Mr Elwes has recently been appointed a director of Kleinwort, Benson Ltd. Until September 1977 he was managing director of Hamilton Brothers Oil and Gas Ltd, which is operator for the Hamilton Brothers North Sea Group. Hamilton Brothers Oil and Gas Ltd developed the Argyll field and has made other significant discoveries in the North Sea, notably the Bruce and Crawford fields. Prior to taking up his appointment with Hamilton Brothers in 1973 Mr Elwes was with the Rio Tinto-Zinc Corporation where he was, amongst other things, policy coordinator for oil, uranium and energy questions, and was also in charge of mining operations in various countries, including Iran (1964/67), South Africa (1968/69) and the UK (1969/70). In 1971 Mr Elwes was seconded to Kleinwort, Benson Ltd to work on various aspects of project finance. From 1972 until 1973 he was head of the RTZ Corporate Division, responsible for finance and treasury matters and financial planning and appraisal. Mr Elwes was trained as a mechanical engineer and before joining RTZ was production controller, Ransomes & Rapier, Ipswich, from 1952 to 1955.

R Kenneth Merkey (Development of Petroleum Financing)

Mr Merkey has recently been appointed president, MC International Inc, Hartford, Conn. He is also consultant to First International Bancshares Inc of Dallas, Texas. Until mid-1977 he was executive director of corporate finance, First International Bancshares Ltd, London, where his responsibilities included project finance, merger and acquisitions and private placements. From 1967 to 1975 he was with First National City Bank concerned with project finance and marketing services to the petroleum industry, initially in New York, later in Singapore and finally in London as vice-president, project finance, Europe and Middle East responsible for arrangements of major project related financings. From 1962 to 1967 Mr Merkey was with the US Navy, Submarine Service; last duty as missile officer on Polaris submarines.

Educated at Yale University, he graduated 1962 BS Industrial Engineering.

George S Miller (Pipelines & Processing Plants)

Mr Miller is a director of Morgan Grenfell & Co Ltd, merchant bankers, London, and is responsible for the bank's business with the international oil industry. From 1962 until 1972 when he took up his present appointment he was with the Continental Oil Company in the USA and Europe, primarily as a professional accountant and latterly in general management. Between 1965 and 1967 he was managing director, Conoco AG — Zurich; in 1967 he was appointed general manager, Conoco Italy and from 1969 to 1972 chief financial officer of Conoco Ltd London. Before joining Conoco Mr Miller was a trainee accountant with Albright & Wilson.

Educated at King's College, Cambridge he holds the degree of MA (Cantab) in foreign languages. He is an associate of the Institute of Cost and Management Accountants and holds the Joint Diploma in Management Accounting.

Otto R Norland (Tankers and Offshore Drilling Rigs)

Mr Norland is executive director of Hambros Bank Ltd London and a director of Ship Mortgage International Bank NV Amsterdam and chairman, Alcoa of Great Britain Ltd, Droitwich. He joined Hambros in 1953 as a trainee, became manager, Norwegian and Shipping Department in 1961 and was appointed to the board of the bank in June 1964. From 1972 to 1976 Mr Norland was a member of the Accepting Houses' Committee and from 1972 to 1975 a member of the British Bankers' Association Executive Committee.

Educated at the Norwegian University College of Economics and Business Administration, Bergen. He was elected fellow, Institute of Bankers in 1970.

Roy T Pleasance (Taxation and the Petroleum Industry)

Mr Pleasance specialises in international corporate taxation. For over twenty years he worked for the British Petroleum Company Ltd, during the last fifteen years of which he headed the taxation department. He was chairman of the United Kingdom Oil Industry Taxation Committee for eleven years starting with its inception in 1964. In that capacity he led all fiscal negotiations with the government on behalf of the oil industry.

Mr Pleasance, who is in public practice, is a fellow of the Institute of Chartered Accountants in England and Wales.

Radu G Plessia (Export Credit Finance)

Mr Plessia is chairman of CIAVE, Paris, and of CIAVE American Corporation USA, and of CIAVE-London Ltd. CIAVE is a financing organisation controlled by Banque Worms, its main shareholder, and Banque de l'Indochine et de Suez. CIAVE's purpose is to help to promote and set up large industrial projects in which multinational export credits are the main financing source. Mr Plessia is also a director of Multi-Credit Corporation of Thailand Ltd, Bangkok. He is a former diplomat.

Born in Romania he is Docteur en Droit et en Sciences Economiques de l'Université de Paris; Diplome de l'Ecole des Sciences Politiques de Paris.

Peter J Wingett (Insurance of Oil and Gas Operations)

Mr Wingett is the underwriter and manager of the Indemnity Marine Assurance Co Ltd London and the assistant group marine manager of Commercial Union Assurance Co Ltd London. He commenced his career with a leading syndicate in the Room at Lloyd's in 1941. After war service as an RAF pilot he resumed his career with the same Lloyd's Syndicate. In 1951 he obtained his appointment with The Indemnity Marine. A specialist in hull insurance he sits on many marine market committees, the most important being that of the Institute of London Underwriters, and the Joint Hull Committee, and he is currently chairman of the London Drilling Rig Committee.

Mr Wingett is an associate of the Chartered Insurance Institute.

Philip Wood (Stock Markets)

Mr Wood is a manager, corporate finance at Hambros Bank Ltd with whom he has been associated since 1969. He is also director, Siebens Oil & Gas (UK) Ltd, a company holding exploration licences in the North Sea. In 1955 he was involved in setting up a company to specialise in dealing in securities in Canada and London and also in managing investment funds. Prior to 1955 he worked for ten years with a member firm of The London Stock Exchange, principally as an investment analyst and also on various rights issues, new issues and related matters.

Educated at the London School of Economics he holds the degree of Msc(Econ) on the subject of the market for long term capital (including the theory and practice of stock exchanges). He is an associate of the Institute of Bankers.

PART A
THE INDUSTRY

The petroleum industry is the largest industry in the world. Its size and international nature stems from three basic factors:
- — petroleum provides a comparatively readily available source of energy;
- — the versatility of petroleum in being able to satisfy a wide variety of energy and other needs;
- — the irregular distribution of petroleum deposits in the earth's crust.

The industry consists of a wide range of activities, spatially separated and technically distinct, but all collaborating in getting petroleum from the ground to the ultimate user. The economic nature of these distinct but complementary activities of the petroleum industry makes it one of the most heavily capitalised of all industries.

Few industries have as high a profile and must cope with the consequent social and political problems. Few industries have to cope with such an array of different kinds of uncertainties. Few industries, therefore, are subject to such a process of progressive adaptation.

In the two chapters which constitute part A the reader is given an overview of the industry and its financing. Chapter 1 describes the nature and structure of the industry and, in particular, emphasises the international dimensions and the factors which contribute to the dynamic nature of the industry. The fact that these factors affect operational and investment decisions which cross national frontiers is what gives rise to many of the uncertainties and risks. Special attention is given to OPEC and its possible future behaviour in respect of crude oil prices.

The combination of the capital intensive nature of the industry and the large and varied risks has resulted in financing innovations of a high order. Chapter 2 reviews the development of petroleum financing and briefly describes some of the ingenuity which the financial community has brought to bear on the industry throughout its history. It concludes that even more sophisticated approaches to financing will have to be evolved in order to accommodate the ever-changing requirements of the industry.

1

The Nature and Structure
of the Industry

Historical development

Petroleum, in one form or another, has been an article of commerce
from earliest recorded time. In Mesopotamia it has been produced
continuously, in the form of bitumen, for nearly 5,000 years.
Seepages of inflammable hydrocarbons, called naphtha by the
Babylonians, were an important ingredient of Greek Fire, a weapon
successfully used for several hundred years for the defence of the
Byzantine Empire.

Distillation, the first step in the refining of crude oil, was employed
by the Arabs and Persians to get lighter fractions for illuminating
oils and by the twelfth century a fairly mature distillation technology
existed in Europe. Some 1,200 barrels of naphtha are said to have
contributed to the conflagration that destroyed Cairo in 1077 and if
this were the case it is evident that a thousand years ago there was
an established petroleum industry in the Middle East.

In the fifteenth and sixteenth centuries oil seepages in Western
Europe were collected and sold for medicinal purposes. In 1712 the
rock asphalt near Neuchâtel in Switzerland was found to provide a
good mastic for floors and steps when heated and mixed with
powdered rock. By the end of the century this material was being
used for roads and in 1832 its properties for this purpose were
improved by adding mineral oil.

All these applications were based on employing seepages of
petroleum that appeared on the surface. The extracting of petroleum,
in the form of crude oil or natural gas, separately or in combination,
from subsurface deposits dates only from the mid-nineteenth century
and as such is a relatively new and specialised branch of the mining
industry. The earlier history gives some indication, however, of the
dispersed location of petroleum and of its versatility as a raw
material, characteristics which, greatly magnified, are typical of the
petroleum industry today.

It was Colonel Edwin Laurentine Drake who was the first to tap
oil underground by drilling. On 27 August 1859, in Titusville,

Pennsylvania, he completed the first well to produce oil. He is, as a result, credited with founding the modern industry. A former railroad conductor, he was employed by the Seneca Oil Company of Connecticut to seek for oil by drilling in Pennsylvania where natural seepages occurred. His employers were so pressed for money that at one stage Drake chose a drilling location which was accessible to him with the pass he had, as a former employee, from the railway. In drilling for oil Drake founded that function of the petroleum industry now known as production, a highly complex branch of petroleum engineering. This development[1] very quickly spread to other countries, notably in the Caucasus in Tsarist Russia where the great oil fields of Baku were found in the 1870s. By the end of the following decade these surpassed for a brief period the output of the USA.

For the next forty years the ever-increasing amounts of oil on the market went, for the most part, to supply an inexpensive illuminant. However it had been apparent from the beginning of the operations in Pennsylvania that petroleum is an invaluable raw material, a fact that was stressed in the report which Professor Benjamin Silliman of Yale prepared for Drake's employers based on samples of Pennsylvanian oil they submitted to him. By the 1890s more than 200 derivatives accounted for at least half of the volume of the industry's total sales.

With the turn of the century and the growing use of the motor car the emphasis became more and more on the production of gasoline, which up to then had been of secondary importance. Ships, too, began to experiment with oil firing. The industry had brought on an age of illumination in the nineteenth century and with the coming of the twentieth it became the moving force behind the age of transportation. The demand for lighter fractions was responsible for the development of the cracking of heavier fractions. Out of these efforts came the treating of oil in a continuous process, first mentioned in the Dubbs patent of 1914 for a process for breaking oil emulsions by heat. The continuous process was later incorporated in the Dubbs cracking process. This was a milestone that presaged the refining process as we know it today. In the 1930s research was intensified on the chemical derivatives of petroleum and a whole new industry was born — the petrochemical industry.

Because of the dynamic nature of its activities the petroleum industry has assumed, particularly since the Second World War, enormous dimensions. Today, it is the world's largest industry. It comprises a wide range of activities, spatially separated and technically distinct, but all collaborating in getting petroleum[2] from the ground to the ultimate user. The main complementary phases of the modern petroleum industry are exploring for and producing petroleum; transporting crude oil to the point of processing or natural gas to regions of use; 'refining' crude oil — that is, converting it into products; transporting the products to regions of use; and finally, distributing them at retail or to the consumer.

The international petroleum industry involves exporting crude oil and/or refined products, and natural gas and/or liquefied natural gas from producing countries and importing them into consuming countries. It is indeed this dichotomy which underlines the 'international' nature of the petroleum industry. Petroleum and its products are commodities in international trade precisely because there are petroleum exporting countries and petroleum importing countries.

Petroleum orientation of countries

Energy is a cardinal factor in the development of any nation, consequently all countries in the world are represented in the petroleum industry as consumers of oil products. However, the vast bulk of oil used in world markets outside the USA and USSR is found in countries whose consumption is currently small and is moved to countries whose consumption is large and by no means matched by any indigenous supplies of oil they may be able to produce.

Fundamental geographical imbalance
The imbalance between production and consumption in the petroleum producing areas is illustrated in table 1.

Table 1: Oil demand in petroleum producing areas (1976)

million barrels/day

Region	Petroleum production	Total oil product demand
Caribbean and South America	3.6	2.6
Western Europe	0.8	13.9
Africa	5.8 ⎫	
Middle East	22.1 ⎬	2.5
Far East and Australasia	2.53	8.0
Total:	34.83	27.0

Source: Adapted from data given in *Shell Information Handbook 1977-78.*

It is therefore convenient to divide countries into three categories in terms of their *principal* involvement in the petroleum industry, namely:

— the large number of countries which absolutely depend on petroleum imports — Germany, France, Italy, Japan, India, etc;

— the steadily increasing number of countries which, whilst being consumers of petroleum, are also absolutely dependent

on petroleum exports. Examples are Venezuela, Saudi Arabia, Kuwait, Iran, Nigeria, Libya, etc;
— the few countries which have a significant degree of self-sufficiency, such as the USA, the USSR, Canada, and more recently Norway and the United Kingdom.

These three categories of country are not mutually exclusive and can change with time. An example is the USA which, before and during the Second World War, was a large exporter to the rest of the world. A turning point was 1947, and since that year the USA has imported increasing quantities from the rest of the world.

The differing circumstances of petroleum exporting countries and the petroleum importing countries has caused the industry from necessity to develop on international lines.

Three zones of the industry
A simplified model of the international petroleum industry therefore consists of three zones — the producing country, the international supply system and the consuming country. An international supply system is a common feature of most raw material industries; however, it has special significance for petroleum and its products, because of the fundamental geographical imbalance between production and consumption in the petroleum producing areas, as discussed previously.

Oil produced in the world outside North America and the Communist countries represented about 60 per cent of world production in 1976. Some 85 per cent of this oil is consumed in countries with inadequate resources for their needs or with no indigenous oil at all. The great bulk — over 50 per cent — of the whole world oil trade therefore spans national boundaries and is currently dependent upon the international oil supply system — and this proportion is continually increasing.

The relevant activities, in a primary vertically integrated sense, for the three 'zones' of the international industry are characterised in terms of physical operations in table 2. Two possibilities are envisaged — one case is with the refinery located in the producing country and the other with it in the importing country.

As soon as one penetrates beyond this rather simple facade, one encounters increasing complexities that make it much more difficult to define or characterise the industry. For instance, the nature of the material being moved — a liquid or gas — necessitates a vast global transportation system of ocean tankers and pipelines, being specially required for the movement of petroleum from where it is found and refined to where it is needed. The petroleum companies are consequently numbered among the largest shipowners and operators

Table 2: Generalised model of the physical facilities comprising the international petroleum industry

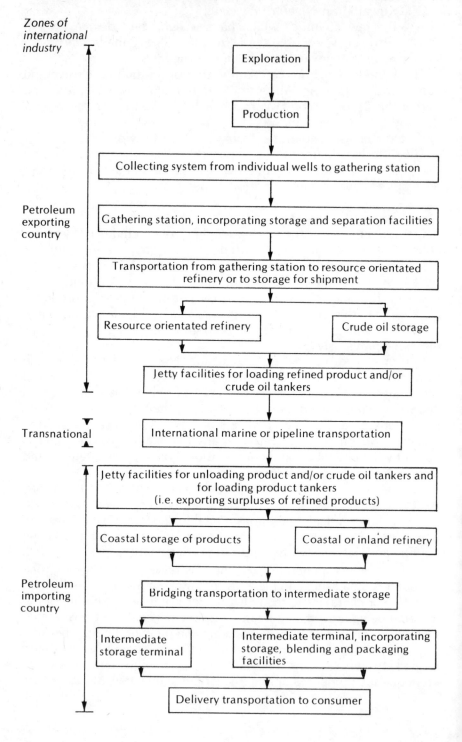

in the world. (See chapter 6 for an outline of the historical development of ocean tankers.)

In the search for oil, companies are inadvertently or contingently involved in the production, use and sale of natural gas, whose processing and channels of distribution are so different from those of oil and its products as to justify regarding it as a distinct, though inextricably related industry. (A brief note on the development of natural gas in international trade is given in chapter 8.) Furthermore, the versatility of the complex mixture of hydrocarbons called 'petroleum' and the progressive development of refining technology have impelled refiners into the broad-scale production of organic chemicals and into chemical processing as well. More than 2,000 products, apart from the familiar fuels and lubricants now issue from the 'petroleum industry'.

To sum up, oil is essentially an international commodity. If countries that have oil are to sell it, and countries that need oil are to obtain it, the oil must be traded across national boundaries. For example, Libya exports crude oil to over twenty countries, while Japan imports over 40 types from over ten countries. Two examples will demonstrate that this is more than a matter of mere commercial advantage. Venezuela depends on oil exports for about 90 per cent of its foreign exchange earnings. Countries of the European Economic Community, on the other hand, have to import about 90 per cent of their oil requirements.

Factors complicating petroleum operations

The international petroleum industry is subject to a greater degree of change than most other natural resource industries. This has resulted in it becoming inherently dynamic in nature and subject to a process of progressive adaptation. Whilst this has been characteristic of the industry throughout its history, it has been especially marked during the time span from, say, the 1930s to the present time due to a number of interrelated developments; not to mention being brought continuously into a state of flux by numerous economic and political events culminating in the progressive takeover of the producing assets of the industry by the petroleum exporting countries, notably in the early 1970s. Some of the more significant developments are the following:

—the enormous growth in petroleum demand, especially in Europe and Japan;

—the decline in relative importance of the Caribbean/US Gulf as a source of supply and the growth in importance of the Middle East and later of North and West Africa;

— the major redistribution in refining activities from producing countries to consuming countries, resulting in a major switch from refined products to crude oil in terms of international movements;

— the increase in the size of tankers from 16,000 dwt to well over 200,000 dwt over this period.

Nature of oil demand

A notable feature of the energy market prior to the 1973 energy crisis had been the progressive substitution of other fuels, particularly coal, by oil. This is dramatically illustrated by the data in table 3. Between 1945 and 1974 the proportion provided by oil fuels increased from 25 to 54 per cent, and that of natural gas from 10 to 18 per cent, and during the same period the share of solid fuels fell from 60 to 19 per cent. Nuclear energy came on the scene in 1960, but is still contributing no more than about 2 per cent. Hydro-electricity has maintained a modest share — 6 per cent — mainly in industrialised countries.

Table 3: Changing pattern of world commercial energy demand[3]

expressed in percentages

	Solid fuels	Oil fuels[1]	Natural gas	Hydro electricity[2]	Nuclear electricity[2]
1920	86	9	2	3	—
1930	75	17	5	3	—
1940	69	21	6	4	—
1945	60	25	10	5	—
1950	52	32	10	6	—
1960	36	41	16	7	0.03
1970	23	52	18	6	1
1974	19	54	18	7	2

[1] Oil fuels include ocean bunkers, aviation fuels and refinery use and loss, but exclude non-energy products.
[2] Nuclear and hydro-electricity are on an input basis.
[3] All data exclude the USSR, Eastern Europe and China.

This means that in a period when overall energy demand has been growing at, say, 5 per cent per annum, oil has been rising by 8 per cent per annum and natural gas also by 8 per cent. Now, following the rapid increases in oil prices in 1973-1974 this process has already shown signs of going into reverse with oil being substituted by coal in electricity generation, etc, and with high taxation on many oil products reducing oil demand still further.

Oil demand is not a monolithic force; in each country it is an aggregation of many parts, each with its own peculiarities and susceptible to the dissimilar, though related forces of geography and social custom, politics and economics, technical achievement, and oil and energy supply. Basically the demand for petroleum results from it providing a comparatively readily convertible source of energy, although an increasing number of products — petroleum chemicals, detergents, waxes, lubricating oils — are being made which fall outside the energy category.

The demand for each of the main energy products, although broadly related to the level of economic activity, is based upon different factors. In the first place, the main products are not generally interchangeable and in most respects must be regarded as entirely different and fulfilling quite distinct functions. For example, motor gasoline (petrol) cannot be used in a portable oil heater, and kerosine (paraffin) cannot be used in a car engine, but can be used in a jet engine.

Secondly, end uses of petroleum products fall into two main categories — 'transport and engine uses' and 'other uses'. In the first category the petroleum industry as a whole has a near monopoly on the supply of fuels (and lubricants). Consequently, the demand for motor gasoline, for instance, depends largely upon the growth of car populations. In the case of the second category of end uses, competition comes from all other sources of energy, for example coal, gas, electricity, nuclear and hydro-electric power, which vary considerably from country to country, as does also the ease with which they can be utilised, and therefore their cost. Hence the demand for fuel oil, for instance, is considerably influenced by availability of other sources of energy and technical developments in industry.

The overall growth in oil demand is a composite of many different growth rates, both within particular countries and by the aggregate oil demand of different countries. And not only do growth rates vary, but in no two countries is the pattern of demand exactly the same. The data in table 4 show, for 1976, how the relative importance of each type of product may vary from country to country. The USA, for example, with the largest car population in the world, takes 45 per cent of the total as gasoline; it takes relatively little fuel oil because it has its own abundant supply of coal and natural gas. In India, 19 per cent of consumption is kerosine, mainly for domestic cooking and lighting. Nearly half the total in Japan is fuel oil for industry and power generation.

Table 4: 1976 Patterns of product demand

expressed in percentages

	USA	UK	Japan	India
Gasolines	45	31	23	18
Kerosines	6	8	9	19
Gas oil/diesel fuel	19	23	13	34
Fuel oil	16	32	46	21
Others	14	6	9	8

Patterns of demand gradually change as economic development brings industrialisation and greater prosperity. Various factors quicken the pace; the expansion of chemicals investment in Japan and in many countries of Western Europe brought a growth in the demand for petroleum feedstocks such as naphtha. Similarly, the switch to jet aircraft by the world's airlines led to a growth in demand for kerosine-type aviation fuels and a corresponding reduction in demand for aviation gasolines. The discovery of new gas fields, coupled with improvements in transportation techniques for natural gas, has made an impact on more and more fuel markets.

There are violent seasonal changes in the pattern of consumption; the degree of fluctuation depends on the severity of the weather and can vary considerably from area to area. For example, in the early 1970s oil companies in North West Europe sold about 5 million tons a month more fuel oil and about 7.5 million tons a month more middle distillates, e.g. gas oils and diesel fuel in the winter than in the summer. Gasoline sales, on the other hand, went up by over 1.25 million tons a month in the summer, and bitumen sales by nearly 1 million tons a month. This requires great flexibility in refining and storage capacity, the planning of supplies and the deployment of tanker tonnage on a world scale.

The physical assets for meeting these demands lie in the variety of source and quality of supplies of crude oil, in the flexibility and range of international means of transport, and in the diversity and spread of refineries all over the world.

Geographical spread of refineries
The refinery can be looked upon as a 'bottleneck' between the crude producer and the ultimate consumer. Hundreds of thousands of wells supply hundreds of refineries which pass their products on to several hundred thousand market outlets with many millions of consumers. Secondly, oil refineries differ from other plants which transform a raw material into usable products; usually the volume or weight of the starting material is reduced in the process of treatment — for instance bauxite/aluminium, iron ore/steel and timber/pulp/ paper — whereas a modern petroleum refinery turns out virtually as much in products as it takes in in crude oil.

These factors, together with the inherent uncertainties in the international oil supply system, lead to a wide range of choice of location and a delicate and quickly changing balance of advantages and drawbacks of locating a refinery at the source of the crude oil or in the vicinity of its markets.

The growth and geographical distribution of world refining capacity outside the USA, Canada, the USSR, Eastern Europe and China is shown in table 5. Whereas the countries excluded from this table are

Table 5: Geographical distribution of refining capacity

Total refining capacity in 1,000 barrels per day

	1939	1951	1960	1965	1970
Europe	367	1,391	3,952	7,739	12,520
Africa	6	23	90	525	794
Middle East	355	983	1,417	1,803	2,259
Far East	198	301	1,168	2,915	5,011
Australasia	4	21	244	504	624
Caribbean	806	1,297	2,356	2,854	3,302
South America	129	272	649	1,063	1,510
Total:	1,865	4,288	9,876	17,403	26,020

Source: Papers published by Petroleum Economics Limited.

all, to some considerable extent, self-contained, the rest of the world consists mostly of inter-communicating areas of oil producer and consumer countries.

The data show that the refining capacity of the area more than doubled both between 1939 and 1951 and between 1951 and 1960. It almost doubled again between 1960 and 1965 and a further increase of 50 per cent took place in the five-year period 1965-70, but there has been a significant reduction in new refinery projects (if any) since 1973.

There has been a clearly distinguishable shift in the geographical distribution of refining capacity. In 1939 refineries in the Caribbean area represented no less than 43 per cent of total capacity outside North America and the Communist Bloc. By 1951 the figure had declined to 30 per cent, by 1960 it was less than one quarter and by 1970 not more than one eighth of refining capacity. The corresponding figures for the Middle East were 19 per cent in 1939, 23 per cent in 1951 and currently just less than 10 per cent. On the other hand, Europe's share of refining capacity went up from 20 per cent before the war to 40 per cent in 1960 and was about 50 per cent in 1970.

More recently, the Far East and Africa have also been developing their refining capacity in line with rising consumption. In particular, the Japanese refining capacity has risen from 640,000 b/d in 1960 to above 3 million b/d in 1970.

Factors influencing decisions on location divide naturally into 'general site factors' and a complementary set of factors which, for convenience, can be referred to as 'global factors'.

General site factors include considerations such as the nature of the site, availability of ample supplies of fresh water, availability and costs of services and possibility of higher construction costs in some countries than in others, etc.

Global factors are concerned with the relative advantages of locating refineries near the source of crude oil, near the consumers to be served, or at some point in between. These depend on a variety of considerations. Among the more important techno-economic and geographic factors are:

 — the relative cost of transporting the oil as crude or as finished products;
 — the relative flexibility of different locations to take advantage of alternative sources of crude, alternative transportation methods, and alternative markets;
 — the economies of size in one large refinery compared with any saving in transportation cost if this capacity is divided among two or more locations.

In specific instances, further considerations may also be important in determining refinery location. For instance, if the world were one economic unit then refinery location would be determined by a reasonably accurate balance of such techno-economic and geographical factors. In reality, the world is not one unit; it consists of a great number of national units, each with a life of its own, giving rise to distinct national policies. Because oil is by volume and value the biggest item in international trade and consequently a significant item in the trade balance of most, if not all, countries, it is not surprising that the oil industry bears the imprint of these national policies. Apart from the direct influence of government policy on location, a factor in the choice of country is the company's assessment of the political risk.

Clearly, some of the factors influencing whether a refinery should be built at one place or another are inherently contradictory. This situation has led some to feel that the problem of optimum location of refineries cannot be solved. Table 6 shows the changing pattern of the three types of refinery locations:

Table 6: Patterns of refinery location

expressed in percentages of total

	1939	1951	1960	1965	1970
Resource (Exports)	69.8	49.7	25.3	16.5	11.5
Intermediate	—	6.5	8.4	6.6	7.1
Consumer	30.2	43.8	66.3	76.9	81.4

These data indicate the sharp trend towards consumer-orientated refineries and away from resource-orientated refineries over the period. There are now clear indications that this trend will be reversed in the future, resulting from the decisions of the OPEC countries to use oil-based manufacturing plants as the nucleus of their plans for industrialisation of their countries. Intermediate refining capacity, of which there was none before the war, reached a peak of about 8.5 per cent of the total in 1960, but its share in total capacity has since declined.

Crude oil not an homogeneous commodity

Crude oil, whilst a very versatile raw material, is far from being an homogeneous commodity. There are heavy, medium and light crudes. There are naphthene-base and paraffin-base crudes and the mixtures of the two. There are high sulphur, low sulphur and sulphur-free crudes. And there are crudes with traces of metallic compounds that make catalytic processing difficult.

Modern refining methods permit a wide measure of flexibility in the use of crude oils and in the range and proportion of products they can be made to yield. Nevertheless, for each type of crude oil there is a certain optimum economic yield of end product. This is rarely in balance with the needs of the local market, and the further a refinery diverges from this optimum, the higher its operating costs.

Neither is it often practicable to change the crude to match the demand. For each refinery there is a limit beyond which it is uneconomic to process a crude oil from outside the range of types for which it was designed, and for all but the most sophisticated refineries this range is comparatively narrow.

Sulphur constitutes a special problem, which has become more significant with the introduction of anti-pollution legislation, since some of the crude oils in most abundant supply — in the Middle East and Venezuela, for example — have the highest content of sulphurous compounds. Sulphur is undesirable in almost any oil product, and refinery processes to reduce the sulphur level of lubricants and distillate fuels have long been a feature of the industry. Under any foreseeable market conditions, however, it would be impossible to recover the high cost of removing sulphur from heavy residual fuels, and the best that can be done with a sulphurous residual is to blend it with a low sulphur product, either a residual from a low sulphur crude or a desulphurised distillate. As discussed earlier, in many countries more residual fuel oil is sold than any other product in the oil trade, and the supply of appropriate crudes for the production of low sulphur fuels presents a complex problem of logistics. There is a further constraint: many low sulphur crudes have one drawback in that they are high pour point[3] crudes and hence cannot in isolation be used economically to produce low sulphur residual fuel oil. A low pour point material is also required to blend down to acceptable pour point limits.

Changing pattern of supply

Table 7 compares proven reserves[4] in 1950, 1965 and 1976, and shows that the position is far from static. The change in the relative positions of the Middle East and North America as sources of supply is one of the features of the modern oil industry. Another has been the growth in importance of Africa, with the development of production in Libya and Nigeria. The pattern is still changing with supplies becoming available from the North Sea and from Alaska. The presently proven reserves of Western Europe account for about 3 per cent of the world's totally proven resources.

Table 7: Petroleum proven reserves

expressed in percentages

	1950	1965	1976
North America	30	10	7
Caribbean and South America	11	6	5
Africa	neg.	8	8
Middle East/Levant	51	62	59
Far East	2	3	3
Communist areas	6	9	14
Others	—	2	4
World total (thousand million barrels)	95	342	579

Source: Shell International Petroleum Company Limited.

As more and more of the world's continental shelves come under the drill, more oil and gas will be discovered and produced offshore. It has been variously estimated that between a quarter and a half of all the new oil reserves may be found under the ocean floor.

Beyond this there is the possibility of further even more fundamental changes as economic means are found of recovering oil from tar sands and shale and — in the longer term — from coal. The operating and investment costs are currently uncompetitive but the potential availability is enormous. One set of investment costs for alternative crude oil sources, using Middle Easten oil as the standard of comparison, is given in table 8.

The Athabasca tar sands in Canada contain more oil than the whole of the present proven reserves of the Middle East, and there is another large area of tar sands in Venezuela. The immense American oil shale deposits in Colorado, Utah and Wyoming alone total three times as much oil again. Brazil also has large deposits. The total reserves of coal are at least fifty times those of conventional crude oils.

Varying costs of producing crude oil
Crude oil production costs vary extensively, according principally to whether the oil field concerned consists predominantly of high yield or low yield wells. Table 9 shows the average daily output per well in the late 1960s/early 1970s in some of the main oil-producing countries.

Table 8: Relative investment costs of various fossil fuels sources

Conventional crude oil:	
— on land (Middle East)	1
— offshore (North Sea)	12-20
Synthetic crude oils:	
— tar sands	25+
— shale	30+
— coal (gas)	30+
— coal (hydrocarbons)	35+

Source: Petroleum Economics Limited.

Other things being equal, it is unprofitable to produce high-cost oil at all. But other things are not equal. High-cost crude may be produced not only for quality reasons, but because there may be logistic advantages that more than compensate for production costs. For example, Venezuelan crude shipped to the USA has a freighting advantage over the Middle East. Moreover, it is generally more economic to produce high-cost oil that has been found than to search for possible low-cost oil. And there is an overall need to diversify sources of supply for security reasons.

Human and natural hazards
The whole intricate balance of supply, transport, refining and marketing is subject to a series of human and natural hazards, sudden and, for the most part, unpredictable.

Wars, revolutions and political upheavals are risks to which a worldwide industry is inevitably exposed on a worldwide scale, and they can call for substantial modifications of supply programmes and plans. The Nigerian civil war and the various wars in the Middle East are particularly dramatic examples in the relatively recent past.

On a less global scale, any stage in the complex process linking oil well with final customer may be interrupted by strike action or sabotage, by accidents such as fires and explosions, and by floods, earthquakes, freeze-ups and other natural disasters. The cost of

Table 9: Average production per well

expressed in barrels per day

USA	17
Venezuela	301
USSR	99
Libya	2,985
Iran	11,838
Kuwait	4,665
Middle East, all countries	4,684
Indonesia	248
World	56

Source: World Oil Magazine.

additional oil storage to provide an insurance against such eventualities is formidable and commercially unviable. Since 1973 some
governments are creating strategic storage capacity to alleviate these
hazards.

Ownership of the industry

There are literally hundreds of separate companies, both large and
small, engaged in the complex operations of the oil business. Many
of these companies operate in only one country but, due to the
nature of the industry, there is a tendency for an increasing number
of them to expand and become interested in some section of the
international trade.

The shape of the international industry has been determined, firstly
by the nature of petroleum — either a liquid or gas. Secondly, by the
fact that it is seldom found in a form in which it can be used, or
where it is needed most. Thirdly, by the inherently dynamic nature
of the industry discussed previously causing it to be subjected to a
process of progressive adaptation. These matters are interrelated and
interdependent and have caused the international industry to be
characterised by two factors since its inception: an extraordinary high
degree of risk and a constant hunger for capital.

These factors are of a much greater order of magnitude than in the
domestic oil industries. This became apparent early in the industry's
life and their impact was illustrated by the rapid reduction over the
years of the number of companies which initially set out hopefully
into the unknown of the international oil industry. As capital requirements grew and risk and uncertainty increased in magnitude,
companies which survived understandably diminished. In this process
seven large groups of oil companies emerged, which are both
international and integrated and are generally listed, *viz:*

    British Petroleum Company Limited
    Gulf Oil Corporation
    Royal Dutch/Shell Group of Companies
    Mobil Oil Corporation
    Exxon Corporation
    Standard Oil Company of California
    Texaco Incorporated

Of the parent companies of these groups, five are American, one
is British and the other, the Royal Dutch/Shell Group of Companies,
is unusual because it has two parent companies of different national-
ities, namely British and Dutch.

Some of these groupings were initiated in the late nineteenth
century; others emerged in the 1920s and 1930s, probably as the
result of the depressed state of the world economy and its impact on

the oil market. Throughout there has been a constantly shifting position in competitive relations with each other and they are constantly facing the competition of newcomers (which has been a feature of the industry's history). This high degree of concentration of the international industry has attracted many studies both by the governments fearing monopolistic tendencies and by independent investigators.[5]

These groups are credited with pioneering the opening up of new oil territory in many countries, for example, Venezuela in the 1920s, new areas of the Middle East in the 1930s, Africa in the 1950s and Alaska and the North Sea in the 1960s. They also claim responsibility for many technological innovations now commonplace in the industry. This technological and other expertise coupled with their financial resources enabled them to expand rapidly after the Second World War to meet the energy demands of post-war reconstruction and development. By 1952 these seven groups of companies accounted for more than 90 per cent of production in the world outside North America and the Communist Centres, 72 per cent of refining output and more than 75 per cent of product sales.

Changing pattern of ownership[6]
This scene changed rapidly as more and more governments set up state oil companies and as private companies — mainly large American companies which had not hitherto operated extensively outside the United States, e.g. Phillips Petroleum and Continental Oil Co. — entered the international oil scene. These developments quickly and radically changed the dominating position hitherto held by the so-called major companies.

By the mid 1960s over 100 state oil companies could be counted. Not all of these were of recent origin, but a new emphasis could be perceived. There was and is, of course, great variety in size, importance and location of these companies. Activities embraced crude production, refining and marketing — and in a few cases these extend to include transportation and the handling of natural gas and chemicals. Another varying factor is the degree of state involvement, whether measured in terms of financial participation or assistance, or political control. The impact of these developments on ownership of the industry is shown in table 10.

State companies are not, of course, the only (even if they are the most striking) manifestation of state involvement. Governments have long been actively concerned with oil through their systems of taxation, licensing and other forms of control, as will be discussed later in this chapter and in chapter 9.

Table 10: Participation of state oil companies in various phases of the petroleum industry

expressed in percentages

	Production	Refining	Marketing
1961	8.6	13.4	10.4
1966	9.2	15.4	12.3
1971	10.2	18.4	14.4
1975	62.5	25.0	21.75

Source: Shell Briefing Service.

The move overseas by the larger American domestic oil companies occurred for at least two main reasons: firstly, the growth in demand for oil products was anticipated to be much greater than in the USA; secondly — and probably the more significant — was that the prospects for new discoveries for oil were seen to be better overseas than in the USA. The latter was quite understandable, bearing in mind that oil exploration in the USA had been going on for a century or more, whereas many countries outside the USA were relatively unexplored. These expectations were realised and large new finds of oil were made. This and other successes further diluted the dominating position of the so-called major companies. However, the share of these independent oil companies in the production phase peaked in the early 1970s and then declined from 1973 onwards as OPEC took over the assets of all foreign oil companies operating in their countries; This is illustrated in table 11:

Table 11: Participation of 'independents' (i.e. non-majors) in various phases of the petroleum industry

expressed in percentages

	Production	Refining	Marketing
1961	7.6	17.4	23.9
1966	11.2	23.0	29.2
1971	12.8	25.3	29.7
1975	7.87	28.8	34.6

Source: Shell Briefing Service.

The changing pattern of ownership of the industry which occurred in the decade or so prior to 1960 and the new discoveries of oil made during this period resulted in surplus crude oil being available, increased competition in the consuming countries and the long downward trend in oil prices. This reduction in oil prices in the consuming countries was the ultimate spur to the formation of OPEC.

Organization of Petroleum Exporting Countries (OPEC)

The initial stimulus to the formation of OPEC came from the reductions in posted prices[7] of 1959 and 1960 in line with the lower realisations being achieved in the markets of the consuming countries. It was formed in September 1960 by Venezuela along with Iran, Iraq, Saudi Arabia and Kuwait with the prime objective of raising the price of crude oil for the benefit of the oil-producing countries. It has since expanded by the accession of Abu Dhabi (now the United Arab Emirates), Libya, Algeria, Nigeria, Indonesia, Qatar, Ecuador and Gabon.[8]

For the first ten years of its existence OPEC was hardly known to outsiders other than a select group of senior executives in the oil industry and a small number of senior government officials in the major industrialised countries. This all changed with the dramatic increase in crude oil prices towards the end of 1973, coupled with the disruption of oil supplies from some Middle East sources. These events shook the world economy to its very foundations and has made OPEC one of the best-known and most important international organisations.

Events leading to the escalation in oil prices

In the early days of the posted price system in the petroleum exporting countries, posted prices were more or less in line with actual prices, but progressively became divorced from actual events in the market-place. This dates mainly from the introduction of the so-called '50/50' profit-sharing agreements which became generally applicable in the early 1950s. The producer-country governments, now that they were directly interested in the difference between production costs and crude oil prices, were understandably concerned about the possibility that, in certain circumstances, the concession-holding company might tend to sell crude oil at unjustifiably low prices to its affiliates. The government position was safeguarded against this possibility by basing the calculation of the 50/50 profit split on the concession-holders' posted price. This posted price — at which the oil is available to all buyers — could not be unduly low, because if it were, the concession-holder would run the risk of having to sell at that price to his competitors.

Around 1957 increased competition in the consuming countries caused the start of a long down-trend in prices. The attempts by the concession-holding companies to pass the price cuts back to the producing countries, through reducing 'posted prices', spurred the formation of OPEC. Thus, from this time posted prices outside the USA retained their relevance merely for the purpose of taxation and became explicitly or implicitly tax reference prices, and the main element in the cost of crude oil — host government take — was protected from the direct influence of the marketplace.

At the beginning of July 1962, the Organization of Petroleum Exporting Countries published resolutions calling for negotiations with the international oil industry to raise crude oil prices and to increase the revenues that producing-governments draw from the proceeds of the oil exported from their countries. The surplus of availability over demand throughout the 1960s prevented the OPEC countries using their strength to do much more than hold the line.

The beginning of the 1970s saw a change in atmosphere. A narrowing balance of supply and demand and the lessening of the surplus-producing capacity that had characterised the 1960s enabled the strength of OPEC, latent since its foundation, to be exercised now to pursue the organisation's aims first in price and then in 'participation' in the existing concessions.

A series of agreements was reached between the companies and member governments of OPEC providing firstly, through the Teheran Agreement, for growing government revenues over a five-year period; secondly, through the Riyadh Agreement, establishing the principles of participation by rising percentages over a ten-year period. These agreements were shaped in such a fashion as to provide a transition over time to new cost levels and a new relationship. They were subverted by the impact of the unilateral actions taken by OPEC from October 1973 onwards.

The combined effect of the various agreements reached by the oil industry with OPEC over the period 1970-73 and the subsequent unilateral actions by OPEC on oil prices and producer-government revenue are summarised in table 12.

Table 12: Increases in posted price and producer government revenues (relates to Arabian Light 34° — OPEC's 'Marker' Crude)

prices and revenues expressed in dollars per barrel

	Posted price	Producer government revenue
1 January 1970	1.80	0.91
14 November 1970		0.98[1]
15 January 1971	2.18	1.27
20 January 1972	2.479	1.45
1 June 1973	2.898	1.82
16 October 1973	5.119	3.45
1 January 1974	11.651	9.31
1 July 1974		9.41[1]
1 October 1974		9.74[1]
1 November 1974	11.251	10.14
1 October 1975	12.376	11.00[2]
1 January 1977	13.000	11.57[2]

[1] Increased revenue results from changes in level of royalties and taxes.
[2] Based on technical costs of 28¢/bbl; all others based on 10¢/bbl.

Apart from the changes in price and revenue, the various OPEC countries during this period also progressively took over the producing assets of the international oil companies in their respective countries. The role of the companies was changed from concession-holders and owners of the producing assets into the providers of technical services and buyers of crude oil.

Implications for the world economy

The price and revenue increases had implications on a much broader basis than the international oil industry. It is now generally accepted that they resulted in at least five permanent changes in the world economy:[9]

— a permanent increase in energy costs;
— a slackening in economic growth;
— a boost to the inflationary spiral;
— an enormous transfer of purchasing power to the OPEC countries and the resulting balance of trade effects;
— structural changes in manufacturing industry, especially energy intensive industries such as metals, etc.

These changes brought about other changes affecting the oil industry. For instance the governments of consuming countries such as Western Europe and Japan, etc increased taxes on many oil products as a means of curtailing consumption and thereby reducing their balance of payments deficits with OPEC countries. These actions by consuming governments, together with those previously taken by OPEC, radically changed the interest of the various parties in the ultimate price of the barrel of products to the consumer as shown in table 13 which is representative of the European situation.

Table 13: Breakdown of a representative barrel of refined oil sold in Europe

expressed in US dollars

	1972	1974	1975
Oil industry: costs and margins	3.05	4.15	5.60
Consumer government revenue	5.60	7.85	10.40
Exporting government revenue	1.75	10.55	11.10
Average proceeds (i.e. consumer price)	10.40	22.55	27.10

Source: Adapted from *The Climate for Oil Investment* (Shell Briefing Service, September 1976).

By 1975 the producing (i.e. exporting) countries took the lion's share and the producing and consuming governments' 'take' together constituted 79 per cent of the cost to the consumer.

The beneficial effects of OPEC action were not only confined to the increased revenues made available to the economies of the individual exporting countries. Perhaps even more important, their actions have dramatically brought to the attention of the world at large how dependent modern society is on energy resources. They have also highlighted the finite nature of most energy resources being consumed at the present time, i.e. oil, coal and gas, thereby emphasising the necessity to devote more attention to energy conservation and to developing means of using renewable energy sources such as solar, tidal, wind, etc before the non-renewable sources are fully consumed.

World oil prices

The level of oil prices is a key factor in considering any investment or financing proposal in the international petroleum industry. Future oil prices can be expected to be largely determined by two factors:
  — unilateral action taken by OPEC;
  — availability of oil and other energy sources from countries
    outside of OPEC countries.
In considering the future level of oil prices it is not unreasonable to look back and ask why 1973 was chosen for a dramatic increase in crude oil prices. Let us therefore consider the difference between 1973 and some of the previous occasions when a major disruption in oil supplies occurred. Take 1951 when Mr Mossadeq deposed the Shah and nationalised the Iranian oil industry; at that time Iran was by far the largest single supply point in the Eastern Hemisphere. Mossadeq's action resulted in the total loss of Iranian crude oil and refined products for four years. Yet, despite this loss of a major source of supply, it only caused minor disruption in world supplies. Alternative supplies were made available more or less immediately, principally from the Arab states and the USA.

A more recent occasion was in 1967; in this instance two disruptions occurred more or less simultaneously. Firstly, the 1967 Arab-Israeli war resulted in the closing of the Suez Canal and the need for supplies to be taken round the much longer Cape route and the consequent scarcity of tankers to maintain supplies. Simultaneously, the Nigerian/Biafran war was in progress with the consequent loss of Nigerian crude oil supplies. Again, as in 1951, the supply problem was relatively short-lived because the USA and other Western Hemisphere producers were able and willing to 'open the taps' and make up the loss in crude oil supplies from the Middle East and Nigeria.

Contrast these situations with that in 1973. Firstly, on this occasion the USA itself was a significant importer of crude oil and therefore was unable to 'open the taps' and make up supplies to other countries. Furthermore, the only significant source of alternative supplies was from the Communist countries. Perhaps it is an over-simplification, but until US crude oil production peaked in the early 1970s any OPEC action to increase oil prices was most unlikely to be effective.

The data in table 3 show that over 70 per cent of world commercial energy demand is met by oil and gas, and the 1973 experience has demonstrated the dominance of OPEC on world oil supplies. To reverse the situation will require time. Alternative oil and other energy resources are available but few can be produced without a lead time of at least ten years.[10]

In summary, while potential demand for fuels other than OPEC oil is high — if only on grounds of security of supplies — there is unlikely to be viable alternatives in terms of quantity until the mid to late 1980s. If programmes are implemented, alternative energy supplies could then become available in progressively increasing quantities, but the costs will be significantly greater than OPEC crude oils. This is, of course, the key to the monpoly power of OPEC. Over a large price range these producers have no fears of competitive substitutes. It naturally follows that the cost of energy in all major importing countries will be largely dictated by OPEC pricing policies.

The future behaviour of OPEC

OPEC is a so-called oil producers' cartel, but has so far not behaved as an ordinary cartel. It has been solely concerned with prices rather than output. There has been no control on aggregate supplies, nor any agreement between members on production shares.

The main strength of OPEC lies in the extreme simplicity of its method of operation. This amounts to the periodic determination of the price of a single type of crude oil — Saudi Arabian light, 34°API — known as the 'marker' or reference crude oil.

Predicting the future behaviour of OPEC is obviously hazardous, since it will depend upon many variables. At the risk of over-simplifying and over-generalising a most complex matter, it is proposed to divide it into two distinct but interrelated parts: ·

- internal, i.e. tensions and stresses within the OPEC cartel;
- external, i.e. developments in countries other than OPEC which could affect the economies of the countries in OPEC.

The *internal factors* can, in the first instance perhaps, best be understood by recognising that OPEC is a disparate group of countries with many dissimilar interests and motivations. These derive from comparative characteristics such as:

population — 32 million in Iran, 7.5 million in Saudi Arabia, 100,000 in Abu Dhabi, etc;

oil reserves — Middle East members have nearly 60 per cent (see table 7) of which Saudi Arabia alone represents over one-third of the reserves;

industrialisation—contribution of manufacturing industry to gross domestic products—20 per cent in Iran; 13 per cent in Algeria; 2 per cent in Saudia Arabia.

Despite these differences there is one item on which their interests are totally at one — namely, the desire to get the best price for the longest period for their wasting asset. Countries in OPEC can be divided into two broad groups in terms of their probable attitude to this objective: those that *need* revenue *now* to sustain industrialisation, etc., e.g. Iran, Algeria; and those that have significant surplus revenues at present production levels, e.g. Saudi Arabia, United Arab Emirates, Kuwait.

Apart from their differing attitudes to oil revenues, the members of OPEC are invariably exercised by the problems of the differences in value of their various crude oils. This is due to the wide range of crude oil qualities available. The problem mainly arises because changes in the aggregate demand for crude oils invariably alters the composition of demand for oil products. Such changes have repercussions on the demand conditions for the different qualities of crude oil and on their relative attractiveness to buyers. It should also be added that a further differential in value arises from the geographical location of the crude oil.

Both of these aspects are inherent in the business and do not arise because of there being a cartel. However, the OPEC mechanism does accentuate the issues. In particular, Saudi Arabia has a unique role because it is the producer of the marker or reference crude. By definition the price of the marker crude cannot be unilaterally altered *vis-à-vis* other qualities or locations so that Saudi Arabia has to absorb the repercussions of price changes for quality and location affected by other producers.

These price differentials are likely to be a persistent irritant to OPEC's cohesion, particularly because of the dynamic and shifting nature of the problems and the difficulties in obtaining meaningful information on value from the marketplace. However, they are unlikely to become critical so long as Saudi Arabia continues to take a long term view of the situation. This country has over 30 per cent of the Middle East reserves and is therefore especially interested in the long term market for oil. In contrast, many other producers have either passed into maturity or are approaching this stage.

The second group of determinants affecting OPEC behaviour are the *external factors,* i.e. happenings in other countries which could have an impact on the economies of OPEC countries. For instance: inflation in industrialised countries; level of world demand for energy and crude oil; alternative energy developments; development of non-OPEC production, e.g. North Sea, Alaska, Russia, China; demand for OPEC crude oils; energy policies of the oil importing nations.

Inflation in the industrialised countries is undoubtedly a key external factor in the short term, because of the direct affect this can have on the import bill of those countries wishing to import capital equipment. At the present time Iran, Algeria and Iraq are probably the countries most vocal in this respect.

Of those factors affecting the demand for OPEC crude oil it can be concluded that the future rates of increase of world energy demand will be lower than in the previous decade due to reduced world economic activity and to energy conservation policies. Furthermore, that oil will not continue to increase its share of the energy market as it has done in the past (see table 3). In so far as alternative supplies of (non-oil) energy were concerned, these will not be available in significant quantities before the middle or latter part of the 1980s.

As a consequence of these factors the indications for the next ten years show an average annual growth in world oil demand nearer to 3 per cent compared with 7-8 per cent before the 1973 crisis. How will demand for OPEC crude oils be affected? — a drop of this magnitude in the rate of growth in demand and hence in oil revenues could clearly support internal tensions seeking price increases for OPEC crude oils.

This drop in growth rate of demand for OPEC crude oils is further aggravated by the availability of new sources of oil. The most notable, of course, being from the North Sea and Alaska. Despite its significance for particular countries, such as North Sea oil to the UK, non-OPEC production is not expected to be material in world terms until at least the late 1980s. On the other hand, all the indications are that the USA will have to import increasing amounts of crude oil for some considerable time and the only sources of such imports are the members of OPEC.

After inflation, the key external factor to OPEC behaviour is undoubtedly the import requirements of the USA. The starting point to understanding the significance of this is to recognise that the American domestic petroleum industry has passed into maturity and is entering decline. Any significant increase in domestic crude oil production has to come either from Alaska or from deep-water exploration or in the form of synthetic crude oils derived from shale oil or tar sands. These are high cost and high risk resources.

The USA has, for many decades, been a country of cheap energy, and, since the 1973 energy crisis, Congress has so far refused to agree a timetable for raising energy prices to world levels. There is no reason to doubt that Congress understands the US energy situation in its fundamental economic realities; equally, there is no doubt it feels that the American people are not yet ready to face these realities. All in all it is hard to see the USA radically adjusting its energy policies before 1980 at the earliest. Few results in the shape of investment in plant and machinery for new energy supplies seem likely to have accrued in the American economy by 1985 or even 1990.

This brief analysis highlights some of the reasons why the national policies of Saudi Arabia and the USA are crucial to the future trend in world oil prices. However, in considering their roles and other factors that are likely to affect the behaviour of OPEC, it is important to recognise the dynamic nature of the situation and especially to appreciate the progressive change which will take place in the internal factors over time, e.g. when Saudi Arabia can absorb all its oil revenues on importing goods for either industrialisation or consumption or both.

What does all this mean in terms of oil prices? Since it is not possible to predict the future, it is necessary to examine a number of separate scenarios.

World oil price scenarios
The possible scenarios are infinite depending on the view taken about the future 'strength' or 'weakness' of the OPEC cartel, i.e. like all organisations there will be 'ebbs' or 'flows' over any period of time; the uncertain availability of alternative energy supplies; the indeterminate nature of government action in importing countries and the level of world economic activity.

Three alternative scenarios will be examined and, for the purpose of the discussion, these have been christened:

'Steady as she goes' — marginal annual increases in world oil prices

'Back to the old days' — drastic fall in world oil prices

'Inflation proof or higher' — world oil prices indexed to rate of inflation (or higher) in industrialised countries

The 'steady as she goes' scenario assumes that OPEC survives broadly as today with internal tensions continuing to be held in equilibrium. This scenario is likely to be favoured by those countries with large oil reserves, but at present only a small industrial base, e.g. Saudi Arabia. These countries do not wish to aggravate inflation in the industrialised countries as this would be detrimental to the value of their own monetary reserves and to the long term development of the oil market. To avoid this happening they are prepared to adjust their own production rather than allow prices to increase too fast. A basic assumption for this scenario is therefore a moderate growth in the world economy, and hence in demand for OPEC oil.

By the mid-1980s the countries favouring this scenario today could have a substantial industrial base if their current plans are vigorously implemented. In these circumstances their interest in preserving the value of their own monetary reserves will be less important to them than today and they could then favour a switch to another scenario.

The 'back to the old days' scenario assumes that OPEC crumbles. Given the vast disparity between Middle East oil production costs and the present price level, the scope for OPEC price reductions is considerable. Such action would make investment in alternative sources of energy grossly uneconomic. It is with this in mind that the International Energy Agency, prompted by the USA, has been working towards some sort of agreement on a floor price for oil, say $7/bbl. Whilst accepting that this is a possible scenario it is difficult to contemplate which group in OPEC would benefit economically from such an arrangement except over a very short period of time. The main motivation for this scenario is major political and/or ideological differences between specific groups or individual countries in OPEC; a remote possibility is a reappearance of the historical rivalries between Iran and Saudi Arabia, but such a development is unlikely to continue for more than a few years — given the relative petroleum reserves of these two countries.

The third scenario — 'inflation proof or higher' — is based on the proposition that either the position of certain elements of OPEC become progressively stronger or demand for OPEC oil is particularly buoyant or both. Any of these would result in oil prices being at least indexed to the rate of inflation in industrialised countries in Western Europe, North America and Japan. This scenario would undoubtedly be favoured by countries such as Iran and Algeria who are using a large proportion of their oil revenues to finance the importation of capital equipment and therefore have a vested interest in at least maintaining the purchasing power of their oil revenues. An important countervailing force to this scenario are the implications it has for other developing countries. This could be catastrophic and is the principal factor mitigating against its adoption, at least to a full extent.

This examination highlights the principal tensions within OPEC over the short term — say up to the early 1980s — namely those countries that wish to maintain the purchasing power of oil revenues on a year on year basis and those that are concerned about the long term value of cash reserves resulting from oil revenues. For a given level of demand for OPEC crude oils, if the industrialised countries brought inflation rates down nearer to historical levels of, say, below 5 per cent, the tension within OPEC is minimised and the 'steady as she goes' scenario is more likely to prevail. If inflation rates creep back to the 'double figure' levels of some years of the 1970s, there is a much greater probability of oil prices following closer to the 'inflation proof' scenario. Similarly, the higher the demand for OPEC crude oil the more probable that the higher price scenario will prevail.

What of the medium term future — say the 1983-90 period? By this time, the oil revenues of an increased number of the member countries will be committed to industrialisation and therefore the balance within OPEC could have significantly changed in this regard. This could result in oil prices moving much closer in line with the 'inflation proof' scenario. In fact, at this stage, should rate of demand for OPEC crude oils continue to increase, e.g. resulting from the failure of the USA to adjust policies, then OPEC countries could well be seeking to extract an even high price. This could be a combination of 'inflation proof' plus a share in the GNP increases in OECD countries.

Only statesmanship on the part of OPEC and their interest in the state of the economies of third world countries would seem to prevent this approach being imposed. Potentially the most important countervailing force to this happening is the availability of the alternative sources of energy. Hence the urgent need for these programmes to be vigorously pursued both in terms of the future well-being of the international petroleum industry and, perhaps more important, in the interests of the future growth in the world economy.

References

[1] Many books have been written on the oil industry and its early pioneers — Rockefeller, Samuel, Deterding, Gulbenkian, d'Arcy and many others — a useful list of references is given in P. H. Frankel, *Essentials of Petroleum* (Frank Cass, 1946).

[2] Although it is often equated with crude oil only, 'petroleum' is a generic term denoting all natural hydrocarbons except those in the coal family. Petroleum so defined is normally found in underground reservoirs as a fluid mixture of chemically distinct hydrocarbon compounds ranging in number from a few to many, but, as delivered to the surface, any mixture is classified as crude oil, natural gas, or a combination of the two. Hence the common practice of using 'petroleum' synonymously with crude oil and natural gas. For convenience where no ambiguity would result, these substances are referred to in the shortened forms 'oil. and 'gas',

[3] Pour point is the lowest temperature at which the oil will pour or flow.

[4] Proven reserves means those quantities of crude oil known with certainty to be present and to be commercially recoverable by existing methods. It is not an estimate of the total quantity of oil which might ultimately be available.

[5] Three such studies are: P. H. Frankel *Essentials of Petroleum; The International Petroleum Cartel* (Report by the US Federal Trade Commission, 1952); and E. T. Penrose *The International Petroleum Industry* (George Allen & Unwin, 1968).

[6] Percentages given in this section relate to the world outside North America and the Communist countries.

[7] The term 'posted price' is of American origin and covers the habit of certain operators to declare — to whom it may concern — at what price they would buy or would sell a certain commodity. It was introduced into the petroleum exporting countries by the producing and refining companies who declared that they would sell at a given price, to buyers generally at ports of shipment, in tanker cargo lots.

[8] There are other 'Third World' petroleum producing countries, e.g. in the Middle East — Neutral Zone, Oman, etc; in North Africa — Egypt, Tunisia; in West Africa — Angola (Cabinda), Zaire, and in the Americas — Mexico, Colombia.

[9] T. M. Rybczynski (ed) *The Economics of the Oil Crisis* (Trade Policy Research Centre, London).

[10] Shorter lead times may be possible in selected countries, e.g. United Kingdom should be self-sufficient in oil supplies from North Sea in a shorter period.

2

The Development of
Petroleum Financing

The oil industry is characterised by some of the most unusual, unique and disparate of features. Because of its capital intensive nature it is one of the most heavily capitalised of all industries. Yet the debt to equity ratio is modest at an average level of 32 per cent. The industry accommodates all sizes of companies and degrees of specialisation. There are extremely speculative enterprises and some very pedestrian aspects of the industry. Few industries have as high a profile and must cope with the consequent social and political problems. Because of its importance it is constantly being barraged with regulations and controls. In spite of the pressures from outside, it remains super-competitive within. Its political leverage is probably the most ineffectual and weakest, relative to its size and importance, of any industry.

Traditional sources of finance

Ever since Colonel Drake discovered oil in Pennsylvania the industry has relied primarily upon equity capital and cash flow from operations to finance its current needs and capital expenditure. Even today the industry is a relatively small borrower in financial markets compared to the size of its capital requirements. In 1975 the 37 largest American oil companies had capital outlays of $25 billion of which $15.4 billion was generated from cash flow and $9.6 billion was raised from outside debt and equity issues.[1] In addition to this, $4.8 billion was paid out in dividends, a remarkable 41 per cent of net income to the shareholders.[2]

The commercial banks did not play a very large role in the oil industry until the mid 1950s. Until then the banks had traditionally made available short term, working capital finance, much the same as was available to any other credit-worthy borrower. Any long term financing was generally considered too risky in this industry. In addition, this tended to be a fairly polarised industry with giants and little fellows and few in between, which meant either one had sufficient cash flow and didn't need term financing or one simply wasn't large enough to obtain such financing.

With the mergers and amalgamations in the USA in the post Second World War period a whole spectrum of oil companies was generated. Producing companies started to integrate downstream and marketing companies were integrating upstream. An active equity market provided adequate capital for the more speculative investments such as exploration. A developing commercial banking industry began to provide long term finance for the safer investments such as refineries and producing oil and gas properties. Out of this period were developed such sophisticated financing techniques as the carved-out production payment and the ABC production payment. Banks were looking for sound, proven lending opportunities and the capital-hungry oil industry provided a stable, constant demand for those funds.

Within the USA larger and longer sources of financing were attracted by the petroleum industry. Cross-country pipelines, with 12 to 15 year payouts and almost cash-register-like incomes, were attractive borrowers for the private placement market, i.e. the large pension funds and insurance companies. By the late 1950s the private placement market was also providing long term funding for production payments. Together with the public debt market, i.e. the bond market, large amounts were being raised for major capital expenditure programmes and for acquisition financing. For many years a 15 to 1 price/earnings ratio was considered the marker rate for major oil companies. The equity markets responded with enthusiasm accommodating the most speculative as well as the most credit-worthy of issues. The financial markets were developed to support the industry, and until recently, rarely was a deal not accomplished because of lack of financial support.

Lending theory
In order to comprehend the development of oil financing it is imperative that one understands and appreciates how lenders have developed over the past 20 to 30 years. It is important to appreciate the implications of lending theory and then it is possible to see the application of this theory to the oil industry.

The first question any lender ever asks is whether the borrower can pay the loan back. There exists a myriad of theories, techniques and hunches in order to answer that question, but almost every decision has a bias. That bias is directed in one of two directions: major consideration is either given to the asset or to the cash flow.

Asset-based lending is the more common and, normally, the more simpler form of lending. A lender must make a judgment on the value of an asset and then determine the percentage of that particular asset to be provided by loans. For demonstration purposes let us consider a most common asset-based loan, the personal house mortgage; it is apparent in most cases that far greater emphasis is placed on the value of the asset than on the income of the individual. In an almost inviolate prearranged formula the ratio of asset value and personal income are combined to determine the amount of loan. With experience, policies at lending institutions can be established so that lending decisions can, and have, become almost rote. Another common form of asset-based lending is the familiar margin loan on securities. Securities of a known value are pledged in order to provide collateral for a loan, normally to purchase additional securities. This collateral provides the lender with a liquid, marketable, valuable asset which can be readily sold to refund the obligation. Taking asset lending to the extreme, probably the most simple form of asset lending is the unsecured, promissory note. This requires a lender to make a judgment on the most elemental of assets, a reputation or a 'name'. Many loans, even today, are based simply on the reputation, creditability or 'strength' of the borrower. In fact, until recent times this was the only criterion for lending.

As lending requests became larger and more complicated the theory of lending had to be enlarged to accept a new criterion — cash flow. An indicative example of cash flow lending is the pipeline financing where repayment of the debt depends solely upon utilisation of the asset and collection of the tariffs. It is not the value of the asset which is important, but the ability of the asset to perform and produce cash.

A good example of changing or upgrading of techniques is the tanker loan. For many years tankers were financed on the basis of the value of the asset or the reputation of the owner. With larger crude carriers and sophisticated, expensive LNG carriers being required, the credit of the ship owners became insufficient. In addition, lenders were unwilling to accept the market value risk of the ship and consequently could not lend more than a small fraction of the cost of the ship. However, as major end-users, such as utilities and major oil companies, required these ships, they were willing to give long term charters to the ship owners. This charter is written in order to be the assurance of cash flow required by most lenders and, hence, VLCCs, ULCCs and LNG carriers are able to be financed because the value of the asset has been reduced to secondary importance and lenders can look to a stream of assured cash flow as the primary credit.

Thus it can be seen that asset lending provides financing which is usually only some percentage of the value of the asset. This is equally true of an office building without leases, or a refinery without crude or markets, or an undeveloped proven ore body. However, once the asset can prove its utilisation and a cash flow can be demonstrated

31

by a lease, a throughput agreement or a charter, a loan sometimes in excess of the present value of the asset can be made available.

This technique of lending based on cash flow not only provides larger financing potential but also provides a phenomenon of great interest to most borrowers known as 'transference of credit'. Small companies have been able to borrow giant amounts using someone else's credit. The ship-owners consequently were the largest and most famous utilisers of this technique. Major oil companies or utilities, of course, had no objection as charters were not normally capitalised, hence there was little or no financial statement impact, and a necessary asset was made available with no investment of capital.

Evolution of new financing methods

The end of the Second World War saw a greatly accelerated activity in the oil industry in the United States. In Texas and Oklahoma thousands of operators, mostly small, single rig companies, were looking for oil. Because of their size and the riskiness of their efforts there was little finance available to them from commercial lenders. The exploration industry existed on equity capital, both their own and private outside investors.

Drilling funds

In order to support the exploration and development activity, a whole new industry, the drilling funds, was formed. This was most welcome as the major oil companies did not have sufficient exploration budgets and the commercial lenders and other financial sources were unwilling to finance these kinds of risks. Initially these drilling funds were provided by private investors to the exploration companies on a very personal, informal basis. There was great incentive for the investor as a large proportion of his investment provided a tax shelter for his income. The laws in the USA provided almost complete dollar-for-dollar tax credit for the investments made in exploration and production. In return for the investment the investor was given a fractional working interest in the wells or leases. Consequently, if oil were discovered the investor became an equity owner of hydrocarbons.

This investment vehicle appealed to most high income bracket individuals as they could obtain complete tax shelter and still enjoy the anticipation in a very interesting, although speculative, investment. The basic philosophy of 'no better, nor worse off' as far as cash was concerned and yet with the opportunity of a bonanza had great appeal. To the operator it meant financial support with only modest control and virtually no influence. The operator was the general partner and made the decisions as to where to drill. The investor always had the right to invest in any specific project, but, as they were generally not knowledgeable in the industry, there was almost complete reliance on the operator.

As the drilling funds gained in popularity, drilling programmes were put together by the larger operators. A partnership was formed with the operator as general partner and the investors as limited partners. The investors, rather than investing in specific projects, were buying shares in a partnership. There were always specific plans in each programme for exploration and development wells, but generally the limited partners invested strictly on the reputation of the operator.

With many successes and still larger failures and frauds, a requirement for a responsible regulated drilling fund industry was developed. At the height of the industry in the late 1960s there were over one billion dollars being raised in public funds and it is estimated that a similar amount was being raised by private funds. The public funds were registered funds regulated by the SEC and on offer to any individual. Stockbrokers as well as trust officers were selling drilling funds. Even several of the major oil companies packaged drilling programmes and sold participations.

Unfortunately, with the change in depletion allowance and tax treatment of high income individuals, the drilling fund lost its appeal as a tax shelter. There still are large amounts being raised by private funds but the day of the public fund is largely over. However, these funds did play a valuable role in giving the domestic oil industry the needed exploration funds to search for oil.

Production payments
Once oil or gas was discovered there was available an asset for attracting commercial finance. In the USA the legislation in each state provides for and permits mortgages or encumbrances on mineral rights and hydrocarbons in place. The lenders thus have an asset which has a market value and, with the improved technology, the amount of reserves could be estimated to the lenders' satisfaction. In fact banks hired engineers and geologists and established petroleum lending departments. With experience, lending criteria and parameters were established and the lending pattern became almost routine. This facility was of great benefit to the small operator and major company alike. The small independent could, in effect, discount his future net revenue from a discovery and raise needed cash for additional drilling. The major oil company could raise needed cash for corporate finance with little impact on the financial statement.

The simple production payment became a basis for many new financing techniques and applications. The basic ingredients of credit were still very much intact: i.e. a secured asset with readily determined market value, a reputable operator with proven technology and experience, and a viable industry and safe sovereign environment. A further key ingredient was the ability to project cash flow accurately and determine the probable pay-out period of the loan. This, then, became an attractive vehicle for lenders and borrowers alike.

The major oil companies during the 1950s in the USA were integrating, amalgamating and merging. The ABC production payment was developed to permit acquisition of producing companies with minimum impact on financial statements and significant tax advantages to the vendor and buyer. The production payment was, in effect, a non-recourse, secured loan to a sterile third party which permitted him to buy an oil future which was then sold on to the major oil company. The basics of the credit were the confidence in the size of the reserves, the ability to produce them, and the ultimate reliance was upon the major oil company to take them. This financial technique was also modified into a carved-out production payment which was used by the major oil companies to generate sufficient revenue to make full use of the depletion allowance in any tax year.

Project financing

This probably was the genesis of what has come to be known as project financing. In the broadest of definitions, project finance is the reliance upon a free-standing project with an isolable and assignable cash flow. In the case of the production payment there was also a high degree of liquidity available to the lender as the mortgage on the assets permitted the lender to liquidate the security in the event of a default or interruption in the cash flow.

The natural extension of project financing was to capital projects, such as the downstream processing plants, transportation systems, and marketing facilities. After the development of a technique for production financing the next major requirement by the industry was for pipeline financing. Pipeline companies were fairly unique borrowers in that they were often owned by several interests; they were regulated by government agencies, the cash flow could be projected fairly precisely, and the pay-out period was rather long. With much cooperation and assistance from everyone involved, giant financing programmes were devised. The commercial banks were generally used for the construction financing. The bond markets, private and public, were used to refinance the construction funds and provide the 15 to 20 year money. Although mortgages were placed on the assets, the basic credit support was the throughput agreement. The pipelines were of benefit to the producer, the offtaker, and of course the operator. The assurance of cash flow in the project required the analysis by the lenders of the reserves of hydrocarbons to be shipped, the ability to ship by the operators and the markets to which the products are intended. The cash flow of the pipeline was not related to or dependent on the value of the hydrocarbons being shipped but was determined solely by a tariff or tolling charge for transportation. Hence, a throughput agreement required either the supplier or offtaker or both to assure the lenders that a tariff would be generated as a result of usage of the pipeline. The throughput agreement became an accepted legal instrument and was used by all varieties of lenders.

The large pipeline financings in Europe also used the concept of the throughput agreement. As larger and more complex pipelines were built involving several countries and many users, this instrument provided lenders with the most convenient method of tying those varied and several obligations together. The use of throughput agreements is continually being enlarged upon. The large pipelines in the Ekofisk system in the North Sea are still being financed with this basic form of project financing.

To digress momentarily, the concept of a *balanced viable* project should be introduced at this point. It will be discussed again later, but to understand project financing further it is important to appreciate this concept. Whether done consciously or not, most lenders prefer and seek balanced and viable projects. *Balanced* means that each interested party is contributing credit support relative to his anticipated individual benefit from the project. *Viable* means that no artificial or contrived support is required in that there exists a market and economically natural place for the project. One partner giving extraordinary support, thereby 'carrying' the other partners or the project, or a partner providing subsidised or conces-sionary support in order to make a project competitive nearly always generates credit problems and can often lead to renegotiation of agreements and contracts at some point of the life of a project. Experienced project financiers have designed agreements which inject necessary discipline into a project and tend to foster balanced and viable projects.

The success of pipeline financing and production payments has never been equalled by any other form of capital project, probably because project financing of processing facilities is necessarily more complex. Mortgages or charges on assets can normally provide security, but to assure cash flow a complex series of contracts must be negotiated. An inter-related series of supply, operating and offtake agreements must provide for all contingencies. A construction or completion agreement is required to ensure successful physical completion of a project. Because of the nature of the industry most refineries were part of an integrated operation and hence were historically financed with corporate funds or guarantees. Until recently crude costs could normally be passed on to the consumer and markets had seldom failed to grow year after year. Embargoes were unknown.

Refineries were built in consuming areas and unless gas could be transported by pipeline it was generally not processed further.

The changing scenario

The oil industry had quite a year in 1973. The power and motive force within the industry was greatly disrupted and re-distributed. The producing countries — OPEC — flexed their muscles and a whole new scenario was generated. Gas would no longer be flared but would be gathered, processed and marketed. Refineries would be built in producing areas. Hydrocarbons would be priced and indexed to a world scale of raw materials, commodities and products. New companies were formed and national companies were given new power and authority to administer the national raw material reserves.

The impact of OPEC decisions

The impacts on the financial community were, and are, immense. The banking system is still adjusting to accommodate the new distribution of wealth and financial requirements. The need for project financing has, very simply, multiplied exponentially. Refineries are no longer built for the traditional $1,000 per daily stream barrel, but maybe four or five times that amount. An LNG project not only involves the processing facility but also the tankers and receiving facility. Projects are no longer being built in developed sophisticated areas with plenty of available labour and infrastructure. Now the cost of a project includes housing, port development, desalinisation systems, railroads, roads and even cemeteries. Politics and sovereign risk exposure are now major considerations for investors and lenders alike.

The enormous size of the investment in new capital projects and the number of projects under consideration has put immense strains on the financial resources of the major companies. In addition, the demand pattern for products has changed so erratically that no longer can assumptions be made reliably on market growth and price elasticity. There is simply very little room for error and an idle facility can literally bring a company to its financial knees.

Never has there been a greater need for balanced, viable projects. Producing countries have the raw materials and a great portion of the wealth; the oil industry has the technology; the consuming countries have the markets. Any major export project must be built and financed with appropriate contributions from each party. Hopefully, the commercial lenders will be able to accommodate the requirements with new and improved project financing techniques. These techniques will need to match and integrate the various sources of funds including export credits, private and public bond markets, commercial bank funds, equity and national funds. The amount of credit and equity contribution will depend upon some judgment by the lenders of the proportional benefit derived by the various partners. In the event that a project must be subsidised or provided with artificial support, the project financing must recognise this lack of viability and must assign the support of the project to the party benefiting from the project in some *pro rata* fashion.

There is, of course, a considerable amount of duplicity in the planning stages for the new projects in the oil industry. Every producing country is making preparations to do *something* with its gas, both associated and natural. Almost everyone has refinery plans. In fact, most developing countries, even those without reserves, have grandiose processing schemes. If every LPG or petrochemical project were built the world would be swamped. There simply is not enough demand over the next few decades to accept that much production. The uncertainty of the market is probably the greatest reservation that project financiers must have.

Even in the most stable and technically sophisticated of countries, the USA, there have been great problems to resolve. Project Independence was formulated at the same time that environmentalists were blocking the Alaskan pipeline. LNG imports have been a subject of much importance by the highest of government agencies. Tax laws were actually changed to provide disincentive in some areas to investment in the industry. The major American oil companies have certainly displayed few common trends: some majors are buying retail stores, some are selling out of Europe, some are concentrating on minerals and nuclear energy, some are centralising while others are decentralising. In Europe there is a new nationalisation wave; every country has a national oil company of some sort, or one in formation, with the common charge of developing equity interests in hydrocarbons.

Future developments
In many countries of the world very elaborate, sophisticated financial support mechanisms have been developed for the oil industry. Project financing has brought the best of commercial bank lending technology to the aid of the industry. In the North Sea there are constantly more new and novel mechanisms being formulated. In Alaska the American private and public institutional markets provided billions of dollars for 25 and 30 years. In Japan the domestic oil companies receive as much institutional support as required.

There still remains a great amount of need. There are many sources of loan capital, especially in Europe, which have yet to be tapped in any significant amount by the petroleum industry. Only the blue-chip, major multinational companies have been able to float bonds and even then only 10 to 15 year maturities are available. With three to five year construction periods and five to ten year amortisation periods, longer term finance must be developed.

The major producing states of the Middle East and North Africa have begun their development of infrastructure and export projects. The mobilisation of these surplus funds, together with outside sources of finance, especially export credits, has stimulated the development of projects.

There remain, however, a host of developing countries with projects to be financed. One of the major problems in the industry ahead is the sharing of sovereign risk. There are many producing countries with lagging oil revenues, burgeoning populations and large plans for development. Lenders, borrowers and investors must impose and exercise a discipline in order to maximise available commercial credits for projects, thereby saving soft loans from surplus funds countries and national funds for infrastructure and non-commercial type projects. It is of paramount importance that developing countries take measures to conserve their sovereign credit. There are large requirements in every one of these countries for social and welfare type expenditures. It seems foolish to use the sovereign credit to finance projects which could otherwise be financed with commercial and private funds. Project financing must be developed to provide the vehicle for the investment of foreign capital and the control of the cash flows of the project. Countries, similarly, must accept the imposition of controls and lending covenants, not only in order to obtain financing, but for their own preservation and conservation of credit.

Even the most experienced and knowledgeable treasurers and bankers are confused by the recent changes in the oil industry. Everyone recognises the need for new financial techniques. Project financing is one answer. There will have to be even more sophisticated developments in order to accommodate the ever-changing requirements. It is going to require the co-operation and imagination of everyone in the industry.

References

[1] *Energy Newsletter* (Citibank N.A., 3 November 1976).
[2] *1975 Financial Analysis of a Group of Petroleum Companies* (Chase Manhattan Bank).

PART B
CORPORATE FINANCING

The international petroleum industry accommodates all sizes of companies and degrees of specialisation. Many of them operate in only one function of the industry, such as refining, production or marketing; others encompass many functions within one corporate entity being, for example, vertically integrated from oil well through to the end consumer, as well as operating in a large number of countries. With such a disparate collection of companies in the industry, it is doubtful that there exists what can be called a purely oil industry approach to corporate finance.

The two chapters in this part consider the philosophy towards corporate financing adopted by the kinds of companies which probably represent the two extremes in the industry. Chapter 3 discusses the practices and organisation of corporate finance followed by the large integrated and multinational oil groups. This illustrates the emphasis placed on cash management by this kind of company, i.e. the importance of internally generated funds as the main source of finance for business development and new projects.

In contrast to the kind of company considered in chapter 3, the approach to corporate financing adopted by the small and single-function company, more often than not engaged in petroleum exploration and production, is considered in chapter 4. This discussion highlights the entrepreneurial outlook of this kind of company and the emphasis which they place on risk taking in the early stages, and then finding ways of raising finance for development, usually through some form of project financing, or alternatively selling out to others if finance cannot be raised.

3

International Integrated Companies

This chapter outlines the way the international integrated companies tackle the problem of finance; whilst there are now many oil companies who are international in the context that they trade in more than one country and are integrated in that they operate in all phases of the business, the chapter will concentrate on how the larger international companies, e.g. the majors, deal with financial matters.

A major might well operate in 50 to 100 different countries with total sales proceeds in excess of £15 billion and net income and annual capital expenditure figures both of the order of £1 billion. These groups are moving large quantities of oil round the world from the oil fields of the Middle East and other production areas to the large consuming markets of Europe, Japan and North America and also to other smaller markets. The history of oil has been that, apart from the few exceptions, oil is rarely found where it is to be consumed and so money as well as oil is constantly flowing round the world — out to the producing areas to pay the host government 'take' either as royalties, taxes or sales proceeds and the costs of production; out to the independent tanker owners to pay the charter hires; and in from the consuming countries as the refining/marketing companies operating there pay for their supplies. It would be common practice within a group to have one company designated as a central trader with a corporate responsibility for this worldwide distribution of oil.

Other international companies meet with some of the financing problems of the oil companies but, generally speaking, these other internationals are not leviathans in international trade, in that their downstream companies are not selling a wholly imported product but a product with a higher indigenous component. The oil companies combine the roles of international investor with international trader.

Finance organisation in international headquarters

The finance organisation at the international headquarters will almost certainly provide for a Controller with responsibility for accounting, audit and financial controls and a Treasurer responsible for all money raising and the collection, remittance, banking and investment of cash.

Most companies organise their international headquarters into line departments with a geographical responsibility and staff departments with a functional responsibility. Companies have been beset by the problem of the division of work between the two and particularly by the question of where the totality of finance staff are best deployed. The line departments are wrestling with the day to day problems put to them by the overseas companies and many of these problems have a financial flavour; additionally those departments are often the channel by which investment proposals make their way to the parent board. A practice is growing in the industry to locate some of the international headquarters finance staff in the line departments where they can make a prompter impact on the decision taking, but they will still refer to the finance function proper for policy guidance.

Taxation and corporate structure

There will be a department for taxation and corporate structure, sometimes as a separate unit and sometimes coming under the Treasurer. If it is separate its work will have to be closely co-ordinated with the treasury. International companies have far too much to lose by evading taxes, but they can and must lawfully arrange their affairs so that double or multiple taxation on their profits is minimised. Important decisions need to be taken as to where a company is to be incorporated, which group company or companies should own its shares, what its capital structure should be, what will be the fiscal impact of dividend declarations and so on, and the international company needs to build up all the expertise it can on these matters in its taxation department (see chapter 9 for further details on this topic).

Insurance

Insurance work is often found located in the Treasurer's Department. It is normal for major oil companies to go in for a significant amount of self insurance, some setting up a wholly owned insurance company for this purpose and some again setting aside part of their annual profits to build up an internal insurance reserve. (A brief outline of the insurance business is provided in chapter 10.) The big companies take this course of action because full market insurance would be extremely costly and, on the law of averages, they could only expect to receive back 40 per cent of their premiums in claims recoveries; this is not to suggest that the insurance world makes vast excessive profits — because history has not shown that they do — but they do have very large administrative costs which of course have to be met from premium income. The oil companies are big enough to take the run of the mill accident internally but they can be, and are, worried about catastrophic losses, and like to try and place these risks in the world's insurance markets. Market insurance is also placed where the services of the profession are particularly worth paying for, for example, the handling of third party claims, particularly motor vehicle third party claims.

Pension funds

The administration of pension funds, too, falls naturally into the Treasurer's Department, both at the international headquarters and in the operating companies. The investment work is not peculiar to an international oil company and calls for no particular comment. However one interesting problem that does arise for an international company operating in many different countries, is whether to go for a funded pension scheme — that is, one entirely separate from the company — or an unfunded scheme where the company itself accrues a pension reserve in its own business. From a pure treasury viewpoint there is a lot to be said for the latter as no valuable cash is dissipated, but often fiscal considerations may suggest the opposite and there is the very important personnel consideration that the employees may have a very definite view one way or the other. Whilst the funded schemes are more common, unfunded schemes exist where French influence has been strong and also in some other European countries.

Management of central funds

The monitoring of group funds is unquestionably the key role of the Group Treasurer and his main tool will be a group source and disposition of funds statement both for the current year and for the years ahead. Ideally he would like to have the information at frequent intervals, but, as collecting and collating the information from several hundred companies is a major task — and there is always pressure to cut down on the burden of the companies to send returns — he will probably have to make do with two or three looks at the current year from just before its start to its end. The statement will show the anticipated funds to be provided from operations, the calls on cash to meet projected capital expenditure, changes in working capital and the likely dividend payments. The very large group companies, which have outside shareholdings, will very likely not be included in this statement on a consolidated basis, but be treated as investments because their financing plans will probably be developed separately from the rest of the group. The Group Treasurer will then look to see what is the shortfall in funds before any fresh money is raised or a run down of central funds envisaged. He will then consider the borrowing opportunities open to him worldwide, the prospects for raising new equity money and also take into account what impact such possible borrowing would have on his group loan/equity ratio; he will bear in mind the possibility of some run-down of his central funds which may have reached a level in excess of requirement for working balance purposes. He will then come to a conclusion as to whether the capital expenditure proposed is both financeable and not to the long term detriment of the group balance sheet and, if he has serious qualms on either score, he will have to recommend to the board a cut back in expenditure.

The importance of cash
Just as in any other business cash is regarded as a vital business asset, but perhaps more so since without proper planning the group's cash resources could so quickly be dissipated unnecessarily round the many companies of the group. The policy therefore is that cash, within the considerable constraints which will be referred to later, should be centralised to the maximum extent. Moreover, despite the seeming attractions of cutting out transactions, the best policy to follow is to bring cash in from the downstream companies to the international headquarters and make the outgoing payments from the international headquarters and hence follow the pattern of operations of the corporate central trader. With operations extending over so many countries and with the risk of factors arising which inhibit the free flow of funds to the international headquarters — wars, civil disturbances, economic crises, all of which may close exchange markets — these major companies all deem it expedient to maintain a healthy level of funds at the international headquarters and this level may be as high as a normal month's inflow of funds.

The quintupling of crude prices in 1973 to 1974 posed a serious financing problem for an international company, because not only did oil stocks and debtors shoot up in value, but there was also the need to build up central funds to a much higher level so as to be adquately protected from the same contingencies as before. It was indeed fortunate for these companies that 1974 was a year of record profit enabling them to finance a high proportion of the increased working capital from retained earnings, as otherwise they would have had to burden themselves with much increased debt.

'Central funds', as they are called, are administered by the international headquarters and one would normally expect to see a sizeable proportion held in the currency of the parent company — American companies would hold in dollars, British Petroleum in sterling, Shell to some extent in sterling. In practice there are only two currencies in which very large sums of cash — and central funds of these groups are likely to be in the range of £500 million to £1,000 million — can be invested to meet the criteria of the companies which, described in more detail later, are security, flexibility and yield; these currencies are sterling and dollars.

The sterling market is, of course, in London, but the dollar market for investment is not confined to New York and since the 1960s has also embraced the flourishing eurodollar market centred in London. Since the Netherlands' market is too small both to accommodate big quantities of short term funds and to provide an exchange market for very large purchases and sales of guilders, the Royal Dutch/Shell Group has operated since 1946 under an agreement between itself and its two host governments that for foreign exchange purposes its headquarters shall be deemed to be in London and that it will subject itself to UK Exchange Control; thus putting itself in a broadly analogous position to BP whilst the existing agreement is in force.

The British Government has on many occasions since the war, and the US Government from 1966 to 1974, restricted the amount of overseas investment that their companies may do. Since working capital also falls within these restrictions, the companies have not had unlimited freedom should they wish to switch their central funds from sterling to dollars, in the case of UK companies, and from dollars to sterling, in the case of American companies. Since the UK based companies both receive large amounts of dollars daily and have to make frequent dollar payments, the UK Government has countenanced their holding a dollar 'float' of cash so as to avoid the need of transferring all money into sterling and then back the next day into dollars; this is not only a convenience to the companies but also makes life easier for the UK Exchange Authorities.

Payments from central funds

Payments to the producing governments are made on specified dates according to agreement, with the take on one month's lifting of oil normally paid 45 days after the end of that month, but there are variants on this and in some cases royalty is paid separately from income tax. Since participation by the host governments in petroleum production commenced in 1973, separate procedures have had to be agreed for the credit terms of the buy-back oil, but here again the practice is to pay for a month's or a quarter's oil in one payment. Payments, when they are made — particularly since the quintupling of prices in 1973-1974, — are very large and cause, for a time, a steep dip in central funds.

At one time all the Gulf States except Saudi Arabia were, for historical reasons, paid in sterling whilst Saudi Arabia was paid 75 per cent in dollars and 25 per cent in sterling. More recently, since oil is now universally priced and traded in dollars, a more logical practice has emerged of paying the host governments in dollars. This switch was not so much due to these governments' reluctance to hold sterling as the wish to cut out the interminable disputes between the companies and the governments as to the dollar/sterling rates applicable to particular transactions, particularly since the advent of floating rates. Since 1973 the use of sterling as the currency vehicle for oil payments has waned considerably and by the end of 1975 only about 7½ per cent of payments for oil were being made in sterling, the balance being paid in dollars.

Receipts into central funds

When the downstream companies remit money to the international headquarters for supplies or services they have, in some cases, the choice of buying dollars (or sterling) or of remitting their own currency for sale into dollars or sterling at the international headquarters; their decision will be guided by the available exchange rates at the time in the various financial centres. Their payments to the international headquarters tend to be at frequent intervals, a little at a time rather than a large one time payment.

Restraints

In the past, at the time before the advent of floating rates, when there used to be reports of big speculation in one currency, the oil companies were often cast as one of the principal villains moving money at will about the world from one currency to another and causing the instability of currencies. The fact that the home governments of the companies subjected them to some control has already been mentioned but there are also other restraints at work.

Whilst, as will be shown later, attempts are made to frame the general financing policy of the group and its companies at the international headquarters, the companies of the group are all separate legal entities — some with outside shareholding — operating under the laws and regulations of many different countries and it is necessary that arrangements between the various companies of the group shall be at arm's length and defensible if attacked. At one time when some overseas activities were carried out simply by branches of British companies and when there was little in the way of foreign exchange regulations — such as was the position at one time in the old sterling area — the overseas branch could make remittances to

London designated simply as surplus cash. However nowadays money can only normally be remitted from overseas in payment for some supply or service, or for repayment of a loan from the international headquarters, or for the declaration of a dividend. Very often all such payments require the approval of the exchange control authorities of the overseas company and approval for anything at all unorthodox may not be forthcoming, particularly when a currency is under attack. So much planning is necessary in deciding how companies are to be financed and there may be little flexibility to make changes.

Investment policy of central funds

The need for large central funds of cash has been stressed, and good and responsible management of such a large mass of cash is clearly of fundamental importance. Once again there is no cut and dried answer because the three principal considerations — flexibility, yield and security — often point to different conclusions. The best possible forecasting of money to be flowing in and out is again of paramount importance, and a good Treasurer must lay on the best information service possible to get news of foreseen money movements.

The out payments are easier to get nearly right, because the large ones are the payments to the producing governments on specified dates and the dividends to shareholders. More uncertainty surrounds the in payments, depending as they do on group profitability, progress in capital expenditure and rise or fall in working capital. Even in the comparatively short term the unexpected is very likely to happen, so the investment of the funds cannot be too rigid, such as may lead to a loss when investments have to be realised in a hurry to be turned into cash.

With all this money going round the system there are, naturally, compelling aims to get money as soon as possible on an interest earning basis and to keep it on this basis until the last conceivable moment. It is best to have one or at the most two banks into which receipts to the central funds will initially flow and to have arrangements with them that unforeseen arrivals of money, particularly those arriving just before close of business, get switched to an interest earnings deposit account. Whilst many banks have a role in lending money or accepting deposits, the funnelling in and out of receipts and payments can only be done efficiently through the company's chosen principal bank.

Since the central funds are not an insignificant portion of the total assets employed in the business — maybe 10-15 per cent — the return gained on them is going to be a material factor in the overall group profit. The larger majors may have funds of £1,000 million, so a one per cent improvement on the yield is £10 million, hence there is a strong incentive to try and enhance the yield. The funds, however, are the group's vital liquid resources and they are of an oil company, not of an investment company. Hence there is certainly no desire or impulse to speculate and out-stretch sagacity in pursuit of yield.

The financial markets capable of absorbing such large quantities of short term money are London for sterling and eurodollars and New York for dollars; the American companies invest most of their money in the New York market and the European majors — Shell and BP — use London, and as the latter are constrained by the British Government on dollar balances, they are very significant investors of sterling, with the dollars that they do retain placed in the eurodollar market. As sterling has weakened against the dollar since the Second World War, the value of cash would have been better protected in dollars, but against this the yield of sterling has consistently been some percentage points above the dollar yield and the extra interest earnings had until 1976 more or less compensated for the sterling depreciation.

A typical sterling pattern of investment to meet the considerations listed above might be a spread amongst the following:
— Treasury bills
— Bank deposits
— Loans to local authorities
— Short-dated government bonds.
The decline in sterling in 1976 seriously depreciated the value of sterling holdings. The Treasury bills provide the desired flexibility in the funds, the security is first class, but normally the yield on them will be the lowest in the funds.

Deposits are only made with first class banks and there will be upper limits of total amounts placed with any one bank. The secondary banks would not be considered as providing the desired security and finance houses would only be used when they are part owned by a major bank. Deposits would be placed frequently to run to a date when a call for a large payment was foreseen.

The local authorities are considered to rank in security only marginally below the central government and attractive deals can often be negotiated with an individual authority, giving a high yield and also flexibility to withdraw funds without penalty after an initial period in a term contract.

Government stock offer excellent security and a satisfactory yield, but if too much is invested there and sales to realise cash have to be made when the gilt-edged market has turned down, then of course losses may be suffered. To minimise the risk of this only stock with a comparatively short life — never more than five years — is bought.

Brokers would be used to arrange the purchase and sale of local authority loans and government bonds. For the buying and selling of

Treasury bills the discount houses would be used and bank deposits would be handled direct with the recipient bank.

Fund investment in the eurodollar market is largely confined to dollar bank deposits and, with all the large international banks operating in London, there is a good demand for these funds. The market in dollar Certificates of Deposit is still a little thin and if any large quantity of stock is put on the market there is a significant price weakening; hence its flexibility advantage loses out to the risk of capital loss disadvantage. In the USA, the indigenous companies use perhaps an even wider range of investment opportunities for the deployment of their central funds, extending in some cases to first class commercial paper. However, they too regard security of investment as a prime factor.

Blocked cash

It has been mentioned earlier the great importance international companies attach to a central mobilisation of cash, so a situation where surplus cash is blocked in an overseas country is viewed with great disfavour. Cash could be blocked because of a misjudged financing policy pursued in the past, for example too high a share capital which would be difficult for a business that has contracted to reduce now, but the situations arise much more frequently where foreign exchange controls have prevented the remittance of local profits. Foreign governments may be reluctant to take such a step as it will obviously tend to inhibit the flow of other inward investment but, nevertheless, in the average year, generally speaking, there will be, at any time, half a dozen instances of such happenings. Here a route of patience may be the only answer, coupled with maximum pressure by the local management on the government, or, more exceptionally, an attempt may be made to fix up a type of barter deal under which an incremental export of some commodity or goods is promoted from the country to a third party who pays cash for it to the oil company international headquarters. Such deals are approached with some trepidation since an oil company management will be moving into what is, to itself, uncharted territory and fingers may get burnt in the process.

Raising of new funds

Throughout the 1950s and 1960s the major oil companies were generally able to finance the capital expenditure that they wanted to do very largely from self-generated funds; profits were not all that large, but, compared with today, the capital expenditure programmes were modest and, with oil prices remaining static, working capital requirements did not change much. Companies from time to time made rights issues and some long term borrowing was done, but, gearing — ratio of long term debt to total capital employed — was in general at a low figure of about 15 per cent.

The 1970s have seen a different picture since due to rampant worldwide inflation, the move into expensive offshore production and companies' yearning for some measure of diversification away from oil trading, capital expenditure costs have rocketed and, with the steep rise in oil prices, oil stocks and debtors have increased substantially in value. The financing of what is considered desirable has necessitated the tapping of fresh money and, with the companies' share prices at a depressed level, equity financing has not been attractive. More loan capital has been needed and gearing has started to go up from the historical 15 per cent to round the 25 per cent mark; even at this figure the ratio compares favourably with the ratios of other industries, but the major oil companies, cognisant that they are operating in an extractive industry with a considerable risk element and extremely anxious to preserve a triple A credit rating, are not willing to see this ratio rise too rapidly or reach a figure which could jeopardise their future ability to raise funds on finest terms.

A company which operates internationally is obviously going to look at the worldwide opportunities open to borrowing. These can be broadly categorised as central borrowing possibilities in the major financial markets of the world — New York, London, the eurodollar market and Switzerland —where generally the credit of the parent company is pledged, and local borrowing opportunities where a local company in any of the many countries where the group operates taps its local market for generally smaller sums and generally on its own credit standing. In those cases when the parent company is pledging its credit, the borrowing may not necessarily be done by the parent company itself but by a finance company, specially created for this purpose, which borrows under the guarantee of the parent. Such a company would most likely be incorporated in a country which gave maximum fiscal advantage both to the lender and to the borrower, that is where there was no withholding tax on interest payments to foreign holders and no income tax on interest receipts from abroad.

Various factors have to be considered in decisions on local borrowing, for instance:

Will the borrowing fit in with the overall financing pattern of the company, that is, will it reduce the need for increased amounts of group finance to be provided or will it be possible for some of the group finance previously provided to be repaid? It would not help the overall group plan if borrowing merely led to a local cash accumulation.

Will the interest cost of the borrowing effectively reduce the Group's tax bill? In some cases, for example the producing operations under old type concessions in the Middle East, interest expense is not deductible for tax purposes and in other cases, due to intense competition or particularly rigid and harsh price control, a local company may be earning no taxable profits.

The strength of the local currency and its chances of appreciating or depreciating in value. It is obviously generally preferable to try to borrow where currencies are weak, but it is often just in these countries that controls on a foreign owned company borrowings are tightest.

Local borrowing opportunities in total in any one year may amount to a worthwhile sum and provide a useful source of funds, but it is unlikely that this alone will be sufficient when a major oil company is looking for sizeable new loan finance. In addition, local borrowing opportunities are often restricted to short term money and a group borrowing programme needs to have a reasonable mix between long and short term money.

A decision on the emphasis between long and short term borrowing rests on the Treasurer's assumptions of the future availability of money and its cost. Since the general trend of long term rates has moved fairly remorselessly upwards now for 20 years, those companies who have made long term issues earlier in the period have done well, although at the time they raised the money it looked like a case of entering into an expensive long term commitment. The companies fully appreciate the different tenure of the two types of borrowing, but in their minds eye they are all borrowing and do not merit the separate balance sheet treatment given to them.

Fund raising from the major financial markets
In the past the markets were London for sterling raising and New York for dollar raising. Both have laid claim in their history to being true international markets, in other words the funds raised could be put to use anywhere in the world. Since the 1960s a new and truly international market — the eurodollar market — arose and this market was the most important one for multinational companies in the years that the Interest Equalization Tax was in existence and there were restrictions on foreigners borrowing in New York and on American companies moving funds from the USA for investment overseas. A few further words on each market.

London sterling. Increasingly balance of payments considerations have forced the British Government to impose a total ban on foreign companies borrowing sterling for investment outside the UK. Moreover, borrowing by foreigners for investment in the UK was limited to 30 per cent of total foreign expenditure, with an obligation to bring in the balance from overseas currencies. Some of the American majors did find long term fund raising by way of debenture or loan stock of some attraction in the 1950s or 1960s when they were building British refining capacity, but more recently with the astronomic rise in British long term interest rates a long term issue at, say, 15 per cent has no appeal. Companies have turned to medium term bank finance tied to variable shortish term interest rates, but such finance is difficult to find in large quantities. Sterling loan finance has, in practice, ceased to be a significant source for the major companies, but the London bankers have built for themselves an important role in the eurodollar market.

New York dollar. In periods of no restrictions by the American Government this is the most popular long term market because a significant sum — $150 million in the early post-war years and $300 million today — can be raised in one issue for 25-30 years; although interest rates have risen in this period from 2-3 per cent to 9 per cent for the triple A companies, money on these terms is still attractive in today's inflationary world. This, of course, has been the natural market for the American majors to use for their long term loan capital and Royal Dutch/Shell used the market in 1948 and twice in the 1960s. In 1975 several American majors had bond issues of $250-$300 million and BP made extensive use of the market to raise funds for its Alaskan investment.

Eurodollar market. This market became of particular attraction to the oil industry in the 1960s and early 1970s when it became difficult to raise money in New York to finance investment outside the USA. Money can be raised more quickly and with far less paperwork than in New York, but the amounts that can be raised in any one issue are smaller and normally the durations of loans are shorter. The eurodollar market is discussed in greater detail in chapter 13.

Continental European Markets. A Swiss franc long term market has existed for some time, but it often means queuing for two or three years and the amount that can be raised is limited to 60-80 million Swiss francs. More recently the Dutch banks have been overfull with guilders stemming from Dutch balance of payments surpluses and they have been happy to lend short/medium term to companies with good names and credit ratings.

Unorthodox fund raising
The oil majors have the choice to operate either directly or through contractors in the various phases of the oil business, starting at the oil well and ending at the car's petrol tank or the householder's heating tanks; if they tried to put their own money into all the investment required in all the phases even the largest companies would be overstretched. A practice has grown of husbanding their own resources,

both financial and human, for those phases of the business upon which direct control is deemed vital and for contracting out in other phases the task of financing assets plus possibly the responsibility for operating these assets.

The movement of oil is a particular phase suitable for contracting out, so other parties are found in the tanker business and in the road — and of course rail — transportation of products. The last phase of the selling business to the retail customer is normally done more suitably by an independent local businessman, although he may need to be provided with loan capital to construct suitable premises. Where it is considered necessary to maintain operational control of the asset, arrangements can be made with another party to provide the capital required and then lease it back to the company. Such assets would not of course be in the company balance sheet, but accounting convention now requires that these transactions should be mentioned in aggregate in a note to the accounts and the interpretation of what should be included for disclosure becomes stricter all the time. Hence companies do not necessarily feel that they are carrying out transactions of fund raising which are not impinging on their balance sheet strength — but they may be gaining in the respect of tapping a source of funds from a body to whom a rental income for fiscal or other reasons is the attractive route for earning revenue on his capital.

Another unorthodox way of raising finance is by way of 'production payments', that is on the security of future oil production, which is really taking advantage of the fact that oil in the ground is, in the case of the large oil companies, not reflected in their balance sheets but still may be good security for the lenders. Chapter 5 is devoted to this means of fund raising, but use of it was confined to North America and so far it has only been extended — and in a limited way — to the North Sea.

Financing of a subsidiary company

Estimates will be available of the projected net capital employed in all of the group's companies, that is net fixed assets plus working capital — and a joint view will need to be taken between the local management and international headquarters as to the breakdown of the funds to be provided over the following categories:

Equity paid-up capital
Retained earnings
Preference capital which might be redeemable
Interest bearing loans from the international headquarters
Non-interest bearing loans from the international headquarters

Supply credit

Outside borrowing

To get the right mix of each source many different considerations have to be taken into account as follows:

(a) Even though the overseas company is a wholly owned subsidiary it has a separate legal entity and the financing pattern, for legal and political reasons, cannot be artificial and hollow. For example, from a narrow parochial view at the international headquarters there could be advantages in having a very nominal share capital and the rest of the finance provided by a loan from the international headquarters repayable on demand, but such an arrangement — even if possible — could be very damaging from other considerations.

(b) If the local currency is suspect there will be obvious merit in examining local borrowing possibilities and where available using these as part of the finance.

(c) Whilst oil companies do not evade tax, they nevertheless do not wish to set up corporate, financial or trading arrangements that expose themselves to paying duplicate taxes, and in all cases they need to consider their tax position both in the country of operation and in their home country. For example, if in the country of the subsidiary company there is a very stiff withholding tax on dividend declarations, or in their home country a similarly stiff tax on dividend income, the oil company would be extremely foolish if it planned to take out all cash surpluses as dividends when it had the alternative to take them out as repayment of a group loan. Tax considerations must play a part in determining the financing pattern.

(d) The oil company will tend to review the return it is getting in the country as the profits made by the subsidiary before paying any interest on a group loan dividend by the total group money invested in the company, but locally it may be viewed very differently. In many cases product selling prices are controlled, and the profit margin fixed by government may be simply a function of share capital or share capital plus retained earnings. Similarly the local trade unions' ideas of profits may be solely related to a return on share capital and they may press exaggerated wage demands because of a misguided view of the company's true profitability. Also when a stage of declaring dividends is reached, from a public relations' angle it is obviously undesirable that the dividend as a percentage of share capital should not be grotesque. These considerations point towards establishing a high share capital element in the group finance provided but, if it is set too high, it may be at the cost of financing flexibility in later years.

(e) As far as the downstream subsidiaries are concerned the parent company will not only be the shareholder, but it, or an associated company, will, in most cases, be the principal supplier of crude oil and products and a supply credit will be provided. Shortening or extending the supply credit is often the most flexible method of increasing or reducing the amount of group finance deployed, although there will be instances where a downstream government intervenes and fixes a minimum amount of credit.

All the above considerations are often throwing up conflicting suggestions of the pattern of the group finance to be followed. It is, of course, vitally important to have cash flow projections as far ahead as possible so as to see whether in the years ahead cash surpluses or cash deficits are going to arise in the subsidiary company; then the answer must be to try and strike a balance between the various local considerations, the tax considerations and the desire to maintain flexibility for the future because the actual future will never be as the forecast.

The above problems are of course compounded when the company to be financed is jointly owned by two or more oil companies, because each company's treasury will tend to view the problem differently and much time and many meetings will be needed to reach a compromise.

4

Independent Oil Companies

One of the interesting features of the oil industry is the way in which very large and very small companies exist and operate side by side. Although, as in other industries, the trend to bigness exists this has not necessarily always been brought about by the absorption of small companies into larger units. Some small companies continue to survive and indeed continue to be brought into existence.

This comment applies primarily to the exploration and production end of the business. The refining and marketing activities of the oil industry represent large and geographically diversified operations involving heavy continuing investment in response to market requirements. As such they are the preserve of the integrated oil companies — the majors and large independents — although small refining and marketing companies do exist and function successfully especially in the United States domestic market.

The financing techniques used in large and small companies within the oil industry differ radically. Large companies are able to finance their continuing activities from internally generated cash flows and the raising of new money by way of borrowings on the corporate credit, with periodical rights issues to fund borrowings and adjust the debt/equity ratio. The tendency is towards a monolithic type of organisation involving close central control and direction of all financing activities. The corporate credit is generally sufficiently good to allow the raising of loan finance on the finest terms. The raising of finance by these large companies on a project basis is comparatively rare, although this practice will no doubt increase since the capital requirements for very large and expensive projects may be too great to be financed by more conventional means.

The aim of this chapter is to examine in more detail the nature of the problems facing smaller companies in financing their activities and the ways in which these problems can be overcome. At the risk of over-generalising, the financial management of small companies calls for flexibility and resourcefulness in dealing with these problems; it might also be said that a measure of opportunism — perhaps better expressed as good timing and judgment — is a necessary ingredient.

The philosophy of the independent exploration company

Oil exploration does not require the mobilisation of large resources. The availability of skilled exploration personnel and access to risk finance are the principal requisites. The scale of the activity will be determined by the successful identification and pursuit of exploration opportunities. There is possibly no professional skill in the world that can exercise so great a leverage on funds invested as that of the exploration geologist. One important discovery can change beyond recognition the prospects of a small exploration company. Indeed it would be fair to say that a number of sizeable oil companies reached their present status in the industry by virtue of a single major discovery. One of the more spectacular recent examples of this proposition is Occidental Oil, whose years of prime growth followed the discovery of major reserves in Libya. We may be witnessing a similar phenomenon in the case of some of the independent exploration companies operating in the North Sea whose exploration successes have earned them interests in major oil fields.

Opportunities for growth based on exploration success create a strong incentive for a small company and its backers. They may, by displaying boldness and originality of thought, win strong acreage positions, especially in new exploration areas. There are some good examples of this in the North Sea, for instance Hamilton Brothers, Ranger Oil, Siebens Oil & Gas, Transworld Petroleum and Pan Ocean Oil.

Because of their simple structure and ease of internal communications, as well as their entrepreneurial character, small exploration companies may be quicker to take new opportunities and more imaginative in their approach to them. They very often, perhaps for these very reasons, attract talent from large companies where the opportunity to display imagination and entrepreneurial flair may not exist to the same extent. Another reason why geologists with a record of exploration success may be attracted to the smaller companies is that there may be a more direct relationship between their success in finding oil and their remuneration than would normally be the case in large companies. Generally speaking the larger companies offer a more secure career, but geologists with a more enterprising temperament may prefer the smaller company environment, where a lesser security of employment may, in their view, be adequately compensated by the material benefits flowing from success in exploration.

Smaller companies will often follow a strategy of acquiring exploration acreage in areas of promise, sometimes where little or no exploration has been carried out. This may be the result of original thinking or of the willingness to devote the necessary time to obtaining concessions from governments or negotiating interests in untested acreage held by others. Again opportunities may be seen in offshore tracts where the water is too deep to permit exploitation based on current technology; in such cases it may be judged that the technology is moving sufficiently rapidly to allow the possibility of development in the near term of any discoveries made.

In the United States it is possible for the smaller companies to obtain potentially productive acreage by bidding at auction for Federal or State leases offered for sale. This has been the means by which Alaskan onshore acreage has been leased and also offshore acreage in the Gulf of Mexico. Bid groups are made up of combinations of large and small companies or, less frequently, small companies without a major. There is inevitably a certain amount of politics involved in the choice of associations for these bid groups; this has been exacerbated by recent Congressional moves to limit what are thought to be harmful associations of major oil companies. Bidding for potentially productive acreage involves the exposure of very large sums of front end money; this financial exposure must be assessed in terms of exploration risk, access to markets for product, availability of development finance and likely profitability of a discovery. In a United States domestic context these parameters can be evaluated to reasonable tolerances and the risks, while they undoubtedly exist, do not discourage the bidding of substantial — and occasionally enormous — sums of money for acreage.

Companies such as Hamilton Brothers have played their part in bids for Gulf Coast offshore acreage for many years, with considerable success. Whilst the experience of individual lease blocks obviously varies, the investment has in overall terms been extremely worthwhile and has been a major contributory factor to the successful growth of the Hamilton Group of companies.

Outside the United States exploration acreage may be acquired by direct negotiation with governments, or, more usually, by application for licences offered for allocation. This will require applicant companies to satisfy the host government of their technical qualifications and operating experience and also their ability to finance exploration and development. The smaller company may face difficulties in showing comparable abilities and strengths to the majors which may also be applicants. However, this is not impossible and in fact a number of smaller companies have been, and are, successful operators in the United Kingdom and Dutch sectors of the North Sea. A more common arrangement will be for smaller companies to associate themselves in groups organised by other, larger companies, of which one will be nominated as operator. Each company will take an interest determined in the light of its assessment of the risks and potential of the area and the likely financial commitment involved in an exploration programme. The company concerned may elect to take an interest large enough to allow some dilution at a later stage by a farm-in, if circumstances should so warrant.

In situations such as those obtaining in the Gulf Coast, where large sums are paid for exploration acreage, the acquisition cost and the subsequent cost of exploration may be substantial in relation to the total investment involved in development. However there is a ready market for oil and gas produced offshore and the industry is already well established, and the geology of the province is known in some detail. In these circumstances it is normally possible to evaluate a discovery to fairly tight limits and to obtain development finance for it. It is often possible to obtain development finance from purchasers by way of advance payments under contract and this subject is dealt with more fully later in the chapter, when development finance is discussed. There have also been examples of funds being put up by consumers to finance exploration. This reflects a situation where there is a shortage of gas to fill the existing transmission system. Finance will often, in these situations, be advanced on a non-interest bearing note secured on reserves and production rates which remain to be proved by drilling. If the criteria set are met then the note will be paid off over the initial productive life of the field, on a unit of production basis. Final maturities are negotiable in individual circumstances, but may be as long as eight to ten years. If the criteria are not met the loan will be repayable, unless other collateral is offered and accepted.

The possibility of gearing up the return on new Gulf Coast discoveries has tended, in recent years, to be reduced by the very large premiums paid for acreage offered. This willingness to pay high premiums for United States offshore acreage has reflected in part the concern of the industry to secure domestic reserves at almost any price in the light of the shift in the balance of power in Middle East oil. As against this, of course, there has been a significant improvement in both oil and gas prices in the domestic market within the last three years, and this has helped balance the equation. The successful opening up of the Gulf Coast has given much encouragement to the offshore potential in other areas of the United States; the offshore East Coast should attract a lot of attention from the industry in the near future. Despite the high front-end cost of this type of exploration activity, it is an attractive field of investment for the smaller company, both because of the inherent value and good security represented by US domestic reserves and the relative ease with which development can be financed.

Financing of exploration activities

Small companies operating in the exploration and production field are normally financed initially on an equity basis. Often these companies have had their origin in privately raised capital. The tax treatment of funds invested in oil exploration in the United States has encouraged the raising of large quantities of risk money. Funds invested in direct ownership of oil properties have frequently been converted into equity interests in companies incorporated to provide a broader base for the raising of further capital and also to establish a form of marketable security by which the original owners can trade their interests or raise further funds by the sale of shares to new investors.

Smaller companies inevitably start from a narrow capital base. The capital base can be broadened over time in a variety of ways and considerable scope for ingenuity exists in this area. In its initial phase an oil exploration company will tend to have its share capital valued on the basis of future prospects, and the price of shares may be traded up vigorously as and when discoveries are made which are thought to be significant for the company's future.[1] This may provide an opportunity to raise further equity funds on attractive terms; however companies financing their activities in this way have to recognise that shareholders investing in future expectations will want to see prospects for a longer term return on their money. This tends to mean not only a sustained rate of growth but, in due course, also the prospect of a flow of dividend income and, possibly, some gearing up of shareholders' funds by borrowing. These requirements imply a growth in cash flow and liquidity which many small companies find hard to achieve. Often the difficulty of reconciling these conflicting demands on available cash flows is such that the owners of a company will conclude that a sale of the business based on a favourable valuation of assets is a more realistic objective. However this is not by any means the general rule and there are many examples of small companies with adequate cash flow backing and borrowing power to finance an expanding business.

Farm-outs
The oil exploration business makes extensive use of the farm-out technique. This, in its most generalised form, involves the trading of an interest in exploration acreage for the assumption by another company of the interest-holder's obligation to drill. In many cases, where there is no obligation to drill, the interest-holding company may simply feel that the risks inherent in the exploration prospect are too high for it to assume the financial burden of further exploration. Essentially the farm-out arrangement allows a company to buy in to

acreage which it judges prospective by the assumption of part of the risk and burden of further exploration. The price will be determined by negotiation in the light of a judgment of the prospectiveness of the acreage; it may take the form of cash, but more probably will take the form of a work obligation on the company acquiring an interest in the acreage. In this way a company owning acreage can retain an interest in any reserves that may be discovered at little or no cost to itself. Likewise another company with no acreage position can acquire an interest in reserves for a cash outlay. The terms of the deal will have been conditioned by the differing views by either party of the geological and other risks involved.

It is precisely because views differ on exploration risks that so much farming-in and farming-out of exploration acreage takes place. Without this simple, time-honoured means of trading off risk, a great many wells might not have been drilled or discoveries made at any particular point in time, if at all. It goes a long way towards explaining the existence of so many successful small exploration companies operating in association with large companies.

Farm-outs covering exploration acreage will not normally involve cash transactions. However if the acreage is regarded as semi-proven or proven there are then a number of variations of the farm-out deal which can be negotiated. A cash consideration is obviously one. Another common arrangement is for the interest-holding company to assign an interest in exchange for the assumption of further drilling costs and agreement to pay an overriding royalty on any future production. This arrangement recognises past expenditure which has gone some way towards proving the existence of a commercial accumulation. Another arrangement that can be negotiated is one where the incoming company will provide support for the financing of development on behalf of the company assigning an interest.

A general objective of the exploration company will be to extend its acreage holdings and the quality of its exploration lands by judicious farming-out of its less prospective acreage and farming-in of acreage held by others which it judges prospective. This requires a blend of exploration and commercial skills operating in close concert.

Drilling funds

In the USA the practice of putting up exploration funds from income-bearing high tax rates is extremely widespread and is an important source of risk money for the exploration industry. This may be done in a number of ways but they are all based on the principle that intangible drilling and other costs are deducted for tax purposes and capitalised for financial statement purposes. Thus abortive expenditure is tax deductible whilst successful exploration expenditure produces a capital asset which may be realisable at a substantial profit. Exploration companies operating domestically and overseas may offer participations in exploration programmes to funds raised from industrial or private investors. This is often done on the basis of an agreed minimum commitment of funds for a period of several years. This arrangement allows the company to plan exploration programmes over a period of years in line with such work commitments as may have been assumed under exploration licence terms or farm-in arrangements. Brokers and investment companies have seen the opportunity presented by this favourable tax treatment of exploration expenditure and have organised drilling funds which attract high tax rate money, which in turn can be invested in favourable exploration prospects.

Illustrative agreements

During the earlier stages of North Sea exploration the desirability of attracting exploration funds of United States origin led to the development of the so-called 'illustrative agreement'. This arrangement allowed an American company to provide finance for a British subsidiary which had been incorporated in order to qualify for the holding of North Sea licences. These qualifications required, amongst other things, that the company be United Kingdom tax-resident and locally managed. The American parent company would, however, require the retention of full freedom to write off expenditure for United States tax purposes, and the concept was therefore developed whereby the American company held the beneficial interest in any production resulting from United Kingdom acreage in exchange for its providing the necessary finance to enable the United Kingdom subsidiary to meet its licence obligations and any financing requirements arising.

The relationship between the American parent and the British subsidiary is defined in the illustrative agreement. This financing arrangement had the tacit agreement of the British Government which recognised the importance of favourable tax treatment of exploration expense incurred by American companies whose participation in North Sea exploration was felt to be of national importance.

The favourable tax treatment of exploration expenditure by the United States Internal Revenue has over the years led to the investment of very large sums of money, often at high risk, in oil exploration. This has been of undoubted benefit to the United States domestic economy and has produced an aggressive oil exploration industry which, from its home bases, has spread its activities throughout the world.

Royalty purchase

A relatively novel approach to financing exploration expenditure has been the development of the concept of royalty purchase. Under this arrangement the prospectiveness of a particular licence or block is assessed prior to the undertaking of exploration drilling. A company specialising in this form of finance will acquire from the licence holder an overriding royalty interest in any future product that may result from drilling. The royalty will probably be graduated to take account of the size and profitability of the discovery; in this way it may be stepped up from a low initial figure to relatively high figures on the basis of increments of production level. This form of finance has not yet been widely adopted but it has the advantage of providing risk finance for drilling without having to raise it through the sale of stock or by way of a sale of interest or farm-out arrangement. A royalty interest also leaves the licence interest unencumbered and the two interests may be — and usually are — dealt in separately.

Financing approach to development and exploitation

For some smaller companies the discovery of oil or gas may be an end in itself. Depending on the company's philosophy and the source of its finance it may be decided to sell the fruits of successful exploration. Alternatively it may be decided to negotiate an assignment of interest or a production-sharing arrangement which would give the discoverer a carried interest in production. Again the company may decide to go ahead and raise the finance for the development of discoveries without any dilution of ownership or reduction of the financing burden.

The decision on how development of a discovery should be handled will depend on a variety of factors. Much will depend upon the company's cash flow situation and credit standing; if the cost of development is likely to be very large in relation to the company's net worth and credit base this will clearly dictate the need for some financing support from a third party. There are a number of examples of this in the North Sea. In other, more developed, oil producing provinces the cost of development may not be very large in relation to a company's resources; in Western Canada, for instance, and in many onshore areas of the United States new oil production can be developed and tied in to existing pipelines for a modest expenditure, although it might be pointed out that in the great majority of such cases the wells are not very prolific and the economics of the production operation would not justify investment on the scale seen elsewhere, particularly offshore. In the major offshore producing area of the Gulf Coast, although the expense of development increases rapidly with increasing water depth, there is a ready market for the production and sophisticated banking support for companies requiring development finance. Much of this finance is advanced on a production loan basis, and this important financing technique is discussed in chapter 5 and summarised below.

Production payment loan

A production payment loan is secured on proven reserves and the collateral security of the owner company is not normally sought, so that the loan does not figure on the balance sheet as a secured charge on the company's assets. The lender will in effect assume the risks involved in the recovery of the amount of oil required to repay the loan. The loan is normally repayable on a unit of production basis or from an agreed proportion of revenue, with final maturity fixed in the light of the field's likely performance. For the lender to accept the risks inherent in producing the oil it is of course necessary for an independent reservoir engineer's report to be prepared and submitted to the financing institution. This report will estimate not only the reserves in place and the percentage recoverable, but will predict the performance on the field on the basis of efficient depletion.

The evaluation of a financing proposal is carried out by groups in which both the financing and the petroleum engineering skills are represented. While the lender will accept the risk on the performance of the reservoir, it will be for the borrower to accept such covenants as may be required by the lender in regard to the efficient development and management of the property. There will also, of course, be covenants to ensure that the title to the property is kept in good standing and is not assigned without approval of the lenders or that other steps are not taken which might diminish the lender's security.

Advance payment method

The gas transmission companies in the USA have developed the practice of providing development finance to secure the commitment of new gas production to their transmission systems. This is an alternative to bank finance in the form of production payment loans. In many cases these loans are interest-free, although the price at which the gas is taken will reflect the financing arrangements. Typically these non interest-bearing advances from gas purchasers are repayable on a monthly basis with payments commencing one month after production begins from the property. The amounts advanced in each case are based on determinations of gas reserves, and there are cases where a proportion of advances made have been repaid following appraisal drilling, where the reserves determined to be in place fall short of what had been predicted and used as the basis for contract quantities. The maturity of these loans can vary between five and ten years.

While this sophisticated approach to the financing of offshore production has greatly assisted the development of the Texas and Louisiana offshore reserves, there are certain difficulties facing producers. Notably there is the problem of determination by the Federal Power Commission of the volume of gas that may be sold on an intra-state as opposed to an inter-state basis. This decision is of considerable significance since much higher prices are available on intra-state gas.

Developments relating to North Sea fields

The production loan concept has not yet reached this point of development in the North Sea. The reasons for this are not far to seek. Firstly, the sums of money required for development are very large in absolute terms and in most cases well beyond the balance sheet possibilities of small companies. The last two years have seen the financing of a number of North Sea fields, some for large companies and others for small companies and others again for consortia of large and small companies. Much ingenuity and effort has been expended in adapting established financing techniques from elsewhere to North Sea conditions. However the general problem faced has always been that the security offered for project finance falls some way short of the bankers' requirements.

Technological and environmental risks

In the absence of previous North Sea oil production history the uncertainty of reservoir performance has been a major impediment to the raising of production loans. Secondly, and perhaps more significantly, the difficulty of operations in the North Sea and the consequent uncertainties of project timing and cost, have been a great deterrent to lenders. Both these areas of uncertainty, whilst serious from the point of view of risk assessment, will undoubtedly become more precisely quantifiable with time. With fields now producing, reservoir performance can be predicted on the basis of experience gained and there will perhaps be less concern on this score in the future.

So far as uncertainty of project timing and cost of development are concerned much has undoubtedly been learned from the first round of field development. Initially the engineering approach to North Sea development was based largely on Gulf Coast practice; well-proven engineering concepts developed in the Gulf were applied to the North Sea without, in all cases, the fullest recognition of the degree of adaptation required to meet differing sea and weather conditions. Construction programmes were developed on the basis of very incomplete weather records and proved to be unduly optimistic. As in all highly capital intensive forms of development, the penalties of programme delay and slippage have been very large increases in capital expenditure and financing costs.

The industry will undoubtedly learn from its experience. The next stage of field development may proceed at a more measured pace. With the increasing realisation that a large number of the North Sea fields already found — and possibly those remaining to be found — are too small to justify massive investment, new production techniques involving lower investment will doubtless be developed. The floating production system in use on the Argyll field may in that sense be a precursor of the future.

Some financing arrangements
The ability to finance the development of new discoveries will directly affect the willingness of companies to risk exploration finance. Nowhere is this truer than in the North Sea. A review of the financing schemes negotiated for North Sea development in the last few years brings out the extraordinary variety of approaches and the ingenuity of both users and producers of finance.[2]

In the case of some at least of the larger companies finance for field development has been raised on a project basis although, by one means or another, the corporate credit of the borrowers has been involved. The two most prominent examples of large field finance have been the Ekofisk and Forties fields. In the case of Ekofisk it was found convenient for the consortium members, all large oil companies, to raise long term fixed interest finance on the basis of individual company guarantees. The finance was however specifically dedicated to Ekofisk development and the debt service arrangements and maturities were based on the predicted field cash flows. In the case of Forties, the credits arranged were again specifically for field development and were structured in the form of a production payment loan. However the requirement for completion guarantees and the arrangements covering the final maturity of the loan have the effect of somewhat modifying the production loan concept in its purest form. The loan was of course very large by any standards and, despite the form it took, there is little doubt that it could only have been arranged for a borrower of the financial strength of BP.

Turning to smaller companies engaged in North Sea field development, the nearest approach to the conventional production payment loan is that adopted for Occidental and Thomson Scottish Associates for the Piper field development. Evaluation was carried out following very extensive appraisal drilling and testing which allowed reserves and reservoir performance to be determined to high limits of dependability. This entailed a considerable 'front-end' investment to obtain the information required by the bankers. The loan is secured on the production of the field following an initial period of production during which the borrowers are required to provide alternative forms of security for the loan. This arrangement has the effect of combining the normal requirement on the part of lenders of project finance for a completion guarantee with the need for field performance to meet certain minimum parameters stipulated by the lenders in the light of the independent reservoir engineer's report. The loans are essentially within the medium term credit bracket and debt service cover is expected to be ample on the basis of predicted costs and sales prices.

A feature of the lending which differentiates it from its American prototype has been the negotiation of an override as a 'sweetener' for the production loan finance; this may be taken to reflect the bank's reading of the borrower's credit standing as well as a recognition of the fact that loan funds provided on a highly geared basis are in effect bearing an equity risk.

Another innovation by the bank providing finance for Piper has been the provision of further credits for the development of the neighbouring Claymore field, under the same ownership. The Claymore field is anticipated to be less profitable than Piper for reasons associated with the reservoir, as well as the more limited reserve position. Loan finance for Claymore has therefore been 'collateralised' by relying in part on the cash flows in Piper; Piper's cash flow will in effect be used as additional backing for service of the Claymore debt and this makes it a much more attractive lending proposition from the banks' point of view. So far as the financing operations for both Piper and Claymore are concerned, it has no doubt been of material benefit to Thomson Scottish Associates to have an experienced operator in Occidental as a fellow borrower on the same general terms. An important area of risk assessment by banks will inevitably be their judgment of the competence of the operator for a particular property to bring the field into profitable operation within the timescale and budgets put forward at the time of negotiation of finance.

Technical and performance guarantees
A review of some of the more recent financings for North Sea development shows clearly the range of possibilities open to the smaller company. Two major trends may be discerned, which will probably play a continuing part in North Sea financing, or indeed financing of very expensive oil developments undertaken elsewhere in the world by smaller companies. One is the practice of obtaining a guarantee from another company to support major borrowings

beyond the limit that banks would regard as prudent for a smaller company. The other is the use of some form of participating interest as a 'sweetener' or inducement to lenders or guarantors.

The provision of a financing umbrella in the form of a guarantee from a major company is an attractively simple method of raising finance for a small company, but it will normally only be feasible if there is a strong incentive to the guarantor. The most likely incentive will be the desire to secure an important long term source of crude oil. This in turn will be influenced by the view taken by major companies of their overall supply position from time to time. For instance, in the immediate aftermath of the Yom Kippur war, much anxiety was felt by the majors over their supplies of Middle East crude, and the possibility existed at that time to obtain finance or financing guarantees from majors for the development of new supplies from politically secure sources such as the North Sea. More recently the supply situation has become much easier and the opportunity for this type of deal has probably therefore diminished, although the recent arrangement between Chevron and Ranger Oil, by which the latter's finance for Ninian has been guaranteed by the former in exchange for Ranger's share of Ninian production, is not without interest.

The other potential source of financial guarantees is the British National Oil Corporation. However it is not easy to discuss this possibility in terms of normally accepted financial criteria because of the difficulty of dissociating the financing activity of the United Kingdom Government and BNOC from their political aims. The professed desire of the Government is to see North Sea oil developed as rapidly as possible and the provision of financing support is certainly relevant in this context. However it is also their aim to secure 'participation' in North Sea fields and it has been found convenient to package the two in any dealings with the industry. Thus the provision of financing support by the Government has become politicised and has to be evaluated in a broad political context rather than as one of a number of financing options.

The form of participating interest adopted in conjunction with the provision of loan finance is normally an overriding royalty. This can be at a flat rate or can be geared to the level of profitability of the operation. There are several examples of this to be found in North Sea financings. For instance, Thomson Scottish will pay their bankers a royalty of 2.5 per cent on the value of crude oil sales from Piper, and a royalty of 3 per cent in respect of Claymore production. It is interesting to speculate whether this arrangement reflects the European parentage of the lending bank, since it has not been normal practice on the part of American banks to seek any form of equity interest as an inducement or added compensation for project loans involving equity-type risks; the more usual course has been to negotiate a risk rate of interest. The philosophy of European bankers has however tended to favour this course, and examples can be found in major mining finance. Another example is Ranger Oil, which will pay a royalty of 8 per cent to Chevron for the latter's financing support for its share of Ninian finance.

Oil Production Stock

Ingenious use of the combination of loan finance and a royalty interest has been made by London & Scottish Marine Oil Company (LSMO) and Scottish Canadian Oil and Transportation Company (SCOT). These two companies are vehicle companies for institutional investors participating in Ninian as members of the Ranger Group. In view of the composition and structure of the two companies it was evidently judged disadvantageous to seek secured debt finance or to dispose of any form of direct interest in the licence as part of their financing plan. This, in effect, limited the financing avenues open to them to the raising of unsecured loans with an equity inducement not taking the form of a direct licence interest. This in turn led to the development of an ingenious and novel form of marketable royalty, described in the SCOT and LSMO prospectus as an Oil Production Stock (OPS). This was offered in conjunction with the unsecured loan on terms which probably resulted in a cost of finance of the order of 17-18 per cent gross, or somewhat more than the cost of a bank finance involving medium/long term credits and a royalty interest. Both the loan stocks and OPS are sterling securities. This is unusual in North Sea financings and will have added significantly to the financing costs. However there were special factors applying in this case. For instance, it would probably suit many of the British institutional investors holding shares in SCOT or LSMO to take up fixed interest paper with high running and redemption yields, and there is also, apparently, quite an active market for the royalty stock. The fixed interest paper and the royalty stock will probably have been taken up initially, to a large extent, by the same investors but will, no doubt, over time flow into different hands in view of the differing characteristics of the two securities. The OPS has been designed in such a way that the tax treatment of payments is optimised for institutional holders and it is likely that this highly marketable form of royalty will come to be widely used in future. It is certainly a good example of the ingenuity being brought to bear by the financial world on the difficult problems of North Sea finance.

Sale of interest

The broad alternative to debt finance is the sale of interest in a property on a basis which will give the vendor, in effect, a carried interest. Such an arrangement was negotiated over the development of the Argyll field. It might be pointed out however that the development of this field was not particularly capital intensive by North Sea standards and whether it comes to be more widely adopted probably depends in some measure on the way in which the smaller North Sea fields are developed in the future. The acquisition of a working interest in a field by a company may appear preferable to the provision of loan finance on any other basis if indeed the risks involved are regarded as equity risks. However, there are a good many precedents for the carried interest type of arrangement in the USA and it may become more widely adopted elsewhere. It has been used in situations where investment in partially depleted fields is required for secondary or tertiary recovery. In such cases the new capital has been recovered as a first charge on revenues and subsequently revenues following pay back have been split on a pre-agreed basis between the original owner and the incoming company providing the finance.

Independents and government influence

This chapter has discussed, in a generalised form, the ways in which smaller companies can finance their activities in both the exploration and the production phases. It has sought to bring out the entre-preneurial nature of the oil business, which is certainly most clearly reflected in the smaller companies which lack the depth of resources of the majors and large independents. The oil industry is the risk-taking industry *par excellence* and much of its vitality results from the free play of competitive forces and the willingness to take calculated risks for commensurate rewards.

It is unhappily the case that the oil industry will have to adapt itself increasingly to live with criteria other than those which it has followed in the past. There is a steady pressure for further involvement in the day to day business of the industry by governments in all oil producing countries. In the developing countries the reasons are to a consider-able extent emotional; the idea of owning a depleting asset which is open to exploitation by foreign interests is highly sensitive and is freely articulated by politicians. The arguments for closer government involvement in the industrialised countries tend to be more sophisti-cated but spring from the same general origins. The desire on the part of the government to control and restrict the activites of the oil industry inevitably complicate the assessment of risk by the industry.

It always seems to be difficult for governments to understand the linkage between present risk and future reward which inevitably has to occupy the minds of those in the extractive industries, whether they be oil or mining. Curtailment of potential reward either by administrative regulation or by tax or, more usually, by both does have a disincentive effect on the pace of exploration activity. Often, however, unfavourable trends such as these are not recognised in a timely fashion by the responsible administrations and there may be significant delays in introducing remedial measures. Unfavourable industry sentiment, brought about by government heavy-handedness, has an immediately deleterious effect on the rate of discovery and the national reserve position and this may take years to rectify. If these generalisations are thought to be too broad it is only necessary to consider the effect that oil price controls have had on the rate of discovery in the United States, or the effect that the tax and royalty 'war' between the State of Alberta and the Federal Canadian Government has had on exploration in Canada.

The United Kingdom
In the United Kingdom it is in the area of government involvement with the oil industry that many of the problems arise. For instance it is plain from ministerial statements that a high priority in government's thinking has been the tax 'take' from North Sea oil. This has led to the imposition of a special tax — the Petroleum Revenue Tax — one of the least satisfactory features of which is that each field is treated separately for tax purposes. This means that the application of cash flow from a producing field to continuing exploration expenditure or to new field development is directly and specifically penalised. The effect of this is that the smaller company, having successfully faced the difficulties in the raising of new finance highlighted above and achieved earnings from North Sea production, is them mulcted of a large proportion of cash flow that would otherwise have been ploughed back into productive investment.

The oil industry is one which takes risks. This serves it well, as it does the economies within which it operates. It is difficult, in the context of a country like Britain, with its low industrial vitality, general lack of investor confidence and shortage of risk capital, to see the logic for specifically penalising risk-taking in so vital an industry as North Sea oil.

A further problem arises from the desire of government to control farm-outs. The embargo on farm-outs over the two year period 1973-75 cannot have helped the rate of discovery, high though this has been. However any serious brake on this activity in the next stage of exploration will be more serious, since the odds on wildcatting successes in the North Sea are getting longer all the time; the prime prospects have been drilled and the prospects remaining to be drilled will tend to be more subtle and higher risk; they will also tend to represent smaller targets, on or near (if not actually on the wrong side of) the economic margin. Risk spreading through the medium of farm-outs will play an increasingly important role in the maintenance of exploration tempo. The longer term consequences of this do not require elaboration.

There has been a view prevalent in government circles that trading in exploration acreage is something to be discouraged. The possibility is seen to exist of licence holders passing on their obligations to others or making a profit from an ownership position granted by government against undertakings by the applicant. This is politically hard to defend, particularly where, as is inevitable under the British system of accountability to Parliament, hindsight is applied. To the extent that there are real dangers in this situation, they would appear to be administratively controllable without the need to impose a general limitation on the industry.

The most significant problem created by the government for the oil industry has been through its wish to acquire an ownership position in North Sea oil and the temptation to use its administrative powers taken under recent legislation to limit the industry's freedom of manoeuvre so as to force it to the negotiating table. This is a dangerous strategy since it creates uncertainty in the minds of industry leaders and their providers of finance; it also impairs the confidence of the industry in the motives of the government, resulting in a reluctance to commit funds to risk situations and leading in turn to a cessation or slowdown in the industry's activities. This is now quite evident in the United Kingdom sector of the North Sea.

The involvement of the government in the oil industry, as in other industries, is probably an irreversible process in the United Kingdom of today, and one must hope that understanding will grow between industry and government as to their respective roles.

References

[1] The shares of exploration companies are not allowed to be quoted on all stock exchanges. For a fuller discussion see chapter 12.
[2] Additional information is given in 'North Sea Finance', *The Banker,* May 1977.

PART C
PROJECT FINANCING

By definition, a **project,** on the basis of unity 'of purpose and economic function, can be considered separately from the activities of the company or group of companies developing it. There are two quite distinct methods of external financing for such projects: the 'gradual borrowing or corporate' approach and the 'project financing' approach.

In the first approach, although the borrower raises the funds for a specific project and indeed may undertake not to use the loan for any other purpose, repayment is not solely dependent on the project's success, i.e. the money is raised basically with the guarantee of the parent company(s) of the participant(s) in the project.

In the project financing approach the financing is so structured that the lenders look to the project itself for repayment, i.e. rather than on the credit of the project owners. The term 'non-recourse finance' is often associated with project finance. Strictly speaking, there will always be recourse to some assets or sources of revenue. The term is generally used to imply that recourse is confined principally to the assets or revenues of the project involved.

In the four chapters of part C, consideration is given to the financing of projects in the four main facets of the industry. Particular attention is given to the assessment of risks and to the approach to security for loan finance used by the banks and other financial institutions.

5

Petroleum Production

This chapter is centrally concerned with the value of petroleum since the origins of the financing of the production of this comprehensive commodity lie more or less at that point in its economic history when the all-important criteria of scarcity and marketability combined to achieve value.

Petroleum production was, and still is, of course, financed significantly by either personal funds or the corporately generated earnings of the major oil companies, major independents or small entrepreneurs. Additionally large proportions of production finance have had to emanate from the banks. The banks, however, entered the scene only when that all-important concept of value was recognised and accepted. It is important to examine briefly the tortuous path that led to that stage of acceptance.

It was principally in the USA at the turn of the nineteenth century that the increasing pace of industrialisation, through intense automation and the increase in world trade with a need for faster and cheaper transportation, inexorably hastened the development of the petroleum industry. The period is marked by frenetic exploration and drilling, resulting in such a spectacular major discovery as the great Lucas well at Spindletop in Texas on 10 January 1901 and culminating in the discovery of the legendary great East Texas field with the Daisy Bradford number 3 well by Joiner and Laster on 5 October 1930. This was oil production on the grand scale and initially it ushered in a period of gross exploitation leading from boom to bust for many in a short time.

The farsighted felt, both for technical and financial reasons, that the uncontrolled exploitation of these major fields would lead ultimately only to the ruin of all. Thus in 1930, five years before the Connally Hot Oil Act was passed bringing conservation to the country at a Federal level, the State of Texas empowered, curiously, the Texas Railroad Commission (TRC) with the responsibility for statewide control of the industry, including that most famous and far-reaching concept of prorationing.[1] This turbulent period has been well documented in a number of books and articles, but it is probably sufficient to say here that prorationing was hardly received with any great enthusiasm by those who felt it would serve only vested interests and would again rob the new breed of wildcat explorer of his new-found wealth. The economic implications however of uncontrolled production were obvious. In 1926 oil was selling for $2.29 per barrel, by early 1931 after Daisy Bradford and the Lou Della Crim, the price had dropped from $1.10 to $0.50 per barrel.

In 1930 the TRC made its first prorationing order and the companies most affected immediately sought and won injunctions against the order on the grounds of marketability. Consequently, by the latter part of 1931, even the price of $0.50 per barrel gave way to $0.15 and finally even dropped as low as $0.06.[2] Martial law was ultimately declared in East Texas to uphold the TRC's authority and for many months the famous battles to prevent the production of hot oil were waged between the military and the ingenious producers.[3]

It was not really until 16 February 1935 when the Connally Hot Oil Act was signed into Federal law and, with the Interstate Oil and Gas Act bringing even more influence and order, that the circle of legal enforcement was finally closed.

From 1935 onwards the oil industry moved largely into a more ordered phase, a phase less colourful, perhaps, but inevitable after the orgy of self-interest marking its heyday.

Early American oil financing

The banks, at first, were predictably reluctant to lend money to finance production. With volatile prices, martial law and fraud identifying conditions in the oil business, there was little incentive to become directly involved. However borrowed funds were necessary to maintain exploration and various inducements were soon to be found. The first and most primitive was through a form of inventory finance. The Dallas banks were the innovators who successfully lent on this basis. It is reported that one of the first direct oil loans made by the Republic National Bank of Dallas was made by a somewhat sceptical banking officer out in East Texas looking into a small field tank of produced crude oil being simultaneously produced and sold to one of the many local refiners.

It was not long before the more adventurous and sophisticated idea of the oil payment was used. (Other approaches such as drilling funds, royalty purchase etc. are discussed in chapter 4.) It was often difficult for the developer to produce sufficient oil with his available funds to induce the bankers to loan on his tangible inventory. Therefore, the procedure emerged of borrowing a sum of money towards the development drilling by pledging in advance two or three times the value of the loan against the oil still to be produced. At first this short term finance was provided by the large oil companies concerned with maximising their oil volumes, although it could be argued that the interests of free marketability were better served for the independent developer if he could finance himself through a bank with no real interest in the oil other than as collateral. Production in Texas was so prolific at that time that the banks increasingly became enchanted with loans that produced returns of considerable magnitude often within months.[4]

Greater technical involvement led many banks to form engineering departments of their own staffed by reservoir engineers, recovery experts, tax lawyers and specialised loan officers who gradually came to deal exclusively with the oil community and who were instantly aware of new finds and early rumours, that so often served in those first days as early warnings where increasingly large sums of money were often at stake.

With the improved technology came the era of the calculated technical risk. The hit and miss era of the early 1930s progressively gave way to the inscrutable logic of the slide rule and relatively precise limits could be set on the available oil collateral for required development funds. By the mid to late 1930s the concept of the development loan, secured only by recoverable reserves, had been thoroughly established.

Between 1946 and 1956 with high allowable days in the post-war boom, oil payment lending reached its peak.[5] By 1956 there were 58,000 producing wells in the USA. However from 1956, with surpluses again being produced, allowables were reduced and again the independents particularly found themselves in a financial squeeze. This period of depressed production however sparked off yet another development in the history of oil financing.

Properties were increasingly put up for sale and usually there were insufficient funds or often no apparent tax benefit available to make the transaction feasible.

ABC oil payment

Again in Texas the initiative was taken through the phenomenon of the ABC oil payment. The ABC was originally utilised in the early 1950s, but it was only in the latter part of the decade that it swept through all the financial institutions of the United States, including the insurance companies who saw in it a secure investment medium. It is estimated that the amount of acquisition financed over roughly the ten year period between 1956 and 1966 was in the order of $3 billion.[6]

The significance of the ABC oil payment lay in its tax advantages and in the non-recourse element of borrowing it provided. It operated as follows; when a property was offered for sale and the purchaser found he probably could not afford the purchase without an element of borrowed funds, and assuming the funds were then borrowed and paid over to the seller, the purchaser of the property immediately began to receive income from the property which was taxable and he additionally had the full and immediate liability on his books for the debt to the bank. For this dilemma the ABC allowed a better tax and balance sheet treatment.

The purchaser would provide, say, 20 per cent of the purchase price. He would then approach a third party company, usually formed for the purpose, who would buy an 80 per cent oil payment interest in the property. This purchase would be financed by a loan from a bank with recourse only to the oil in the ground. The seller then received from these two simultaneous buyers his total price. What was the benefit of this apparently tortuous financial transaction? The original purchaser was able, for barely 20 per cent of the price, to purchase his property. His 20 per cent, which was a working interest, enabled him to take full depreciation on his tangible drilling and operational expenses so that his 20 per cent of taxable income was sheltered. At the same time an 80 per cent non-recourse loan was paying off to the bank, for which he had no liability should anything go wrong. The 80 per cent oil payment being attached to no other income could obtain 100 per cent shelter on the principal income under the cost depletion allowance of the time.

However, by 1969 the United States Internal Revenue Service decided that this method of finance was allowing valuable tax income to escape through the tax net and the tax allowance on the oil payment, except for development costs, was summarily disallowed.

It is important at this juncture to stress that our concentration on the history and development of technical oil finance has focused exclusively on the United States as it never appears to have been significantly utilised or documented elsewhere. This, one may suppose, is partly because of unattractive tax possibilities and partly because there was never a sufficient volume of domestic oil discovered, say, in Western Europe, to generate the creative levels of financing we have described in the United States. Furthermore, nowhere but the USA has ever really enjoyed the mass phenomenon of the small independent oilman, the ultimate maverick in the history of the petroleum industry and probably responsible for the discovery in the Western hemisphere of more oil than the combined forces of all the international majors. It was this maverick explorer who was continually short of money and who continuously evoked, and often provoked, the banks into providing for his unique and complex financial needs.

Evaluations for non-recourse loans

Banks are typically approached to lend money against two basic types of development requirements. The first usually involves a new property on which a number of exploratory wells have been drilled, indicating presumptively the presence of oil. The second type is normally encountered in a mature oil province in which the loan is sought to finance undeveloped acreage with collateralisation through the pledge or mortgage of previously developed producing property.

These loans are secured on proven reserves and the collateral security of the owner company is not normally sought, so that the loan does not figure as a secured charge on the company's assets. The origins of this form of finance were American as we have seen, although time has wrought a considerable evolution. These propositions demand distinct forms of analyses from the banks in arriving at a loan decision.

Reservoir analysis

In the case in which there is basically a virgin field with no production history, the reservoir analysis will depend on what is conventionally referred to in the profession as the Volumetric approach.[7] This involves three basic estimates, namely, an estimation of the volume of net productive rock, estimation of the recovery per unit volume of net productive rock and, finally, multiplication of the volume of productive rock by the recovery per unit volume. These data are fundamentally obtained from interpretation of the electric or radio-activity logs and available core samples taken from the various exploratory wells in the reservoir. From this basic information the engineer will build up his estimation of the volume of productive rock in the reservoir known as acre-feet. One acre-foot contains notionally a constant amount of barrels of oil. By a series of eliminating calculations he reduces this constant by his estimate of factors, such as:

— porosity (the volume actually occupied by the oil);
— connate water;
— the shrinkage factor measuring the change in volume of oil between the reservoir and the surface;
— the so-called permeability factor which is a measure of the recovery possibility of the reservoir.

An alternative method of analysis is used when the property in question has achieved some level of performance from which the engineer is able to deduce certain features which will fairly decisively provide him with the picture of viability or not upon which a loan value must be predicated.

A producing field will be revealing against four basic questions:
- Has there been a decline in the production capacity of the wells?
- Has there been a decline in reservoir pressure?
- Has there been a measure of encroachment of water into the reservoir?
- Have there been any sequential changes in producing gas/oil ratios?

It seems to be generally agreed amongst reservoir engineers that the Performance method is the most reliable, although this must be qualified by the caution that volumetric estimation increases in accuracy in proportion to the homogeneity of the sands in a reservoir. There have been remarkably accurate volumetric estimates in the history of oil field analysis. Most engineers, however, will usually be found to estimate extremely conservatively in complex reservoirs where they have nothing but volumetric data to rely upon.[8] Nowhere must this as much be the case as for the banking engineer whose reserve estimates will largely govern the appropriate loan values ultimately agreed upon.

Bank safety factor
With the reservoir analysis completed the decision on loan value must be taken. The banker is now confronted with a field which is projected to produce at a certain rate over a certain estimated interval of time. Depending on the degree of caution in the advice from his engineer, he will decide on what is known as a safety factor, which is basically the number of times he would like his loan covered by the financial equivalent of the expected production of the field after all expenses are met, including taxes. Typically, this figure is usually between 2:1 and, for an extremely cautious estimate in complex reservoir types, 2.5:1. If 2:1 is the adopted safety factor the loan value is 50 per cent of the projected net revenues discounted by the prevailing interest rates. The second decision is now to assess the production profile of the field and decide what percentage of the revenues should be applied to debt service over the term of the loan. For a short life field this might require 80 per cent of the gross proceeds applied to the loan, or again, more typically, it can be lower and is often geared to a mutually acceptable loan package between the banker and his borrower.

Security for loan
If the customer accepts these findings and needs the arrangement on a non-recourse basis he will assign his complete interest to the bank for the duration of the loan. In the United States — because mineral title is personally held — this involves the banks in searching and

establishing true title to the property in question and then filing a mortgage on the property. In the event of a default by the borrower the bank will foreclose its mortgage and have the field sold or produced to satisfy the loan. The situation concerning title to the oil or gas in the ground tends to be more complex in most other countries and this is discussed in a later section.

In domestic onshore oil financing in the United States it was the practice in many banks throughout the life of these non-recourse loans, if they were extended or renewed, to police continually the wells with travelling petroleum reservoir inspectors from the banks. These men would often travel several thousands of miles each year testing wells for pressure and production readings to confirm that the true collateral was being accurately calculated and reported. Clearly, as production has become more remote and exotic in areas such as Alaska and the North Sea, this practice is largely disappearing and the banks have progressively come to rely on the companies to accurately report their production data, the accuracy of which is usually insured by the banks with rather stringent loan default provisions.

Operational environment

Apart from the inherent risks associated with petroleum reservoirs, the banker must also take into account other risks associated with the project including the climatic conditions or environment in which the operations are being carried out.

North America has been somewhat fortunate technically in having a vast proportion of its petroleum reserves exploitable from land drilling. It has only been relatively recently that the offshore possibilities of the Mexican Gulf and the Californian coasts have produced a further technical dimension. However, for the most part the American offshore operation has not been environmentally severe and as such has not significantly altered the perspective of the banks with regard to security.

In contrast to this, a *sine qua non* to project lending in certain difficult international offshore petroleum developments is the use of the so-called technical or completion guarantee. As a component of production lending, it has, in company with other historical developments, been the product of the pragmatic and the expedient to deal with certain specific problems. In most cases it has been provoked largely as a result of the climatic or environmental hostility of the area, giving rise as it does to new and untested technology with the attendant uncertainties of project timing and cost.

The resulting technical guarantee has been described as a form of performance bond, which it superficially resembles. It was first used in a major petroleum loan when British Petroleum sought, from the banks, finance for its major Forties field discovery in the North Sea located deep out in an extremely hostile climatic environment (further details of the financing arrangements for the Forties field are given in Chapter 13). The problem of coping with the environment and engineering a feasible development scheme that could be implemented within the intended time frame was of immense concern to the banks, particularly when the borrower was looking for a non-recourse or limited non-recourse type loan. The technical guarantee became the solution to this dilemma. In exchange for persuading the banks to accept a certain level of risk that the oil would, in fact, be where it was supposed to be, BP undertook to guarantee the technical or engineering completion of the field.

Additionally, however, the need to reinforce the security of the loan has forced the oil banker to go even further. The judgment must be made that, regardless of the technical guarantee which effectively only covers phase one of the loan, the loan value itself must be heavily discounted in favour of the continuing sense of technical hazard. It is for this reason that the safety factors — namely the ratio of the loan to the net revenues expected — have been set as high as a 2.5:1 range in the North Sea. The banker must rely upon this level of safety cover in assessing the overall value of the investment to assure himself that, in the event of a disaster, he has loaned against a valuable field. A valuable field will either be reinstated or will have a resale value sufficient to ensure payout.

Financial standing of borrowers
The North Sea has produced a few other novel complexities, perhaps in some respects less onerous and disturbing than the climate, but in other ways more difficult to deal with legally or financially.[9]

One of these novelties was the form of allocation of licences. In the first flush of national euphoria it was decided to allocate licences to a variety of applicants. The production licences for the United Kingdom sector were issued to an array of companies ranging from multinational major oil companies to major and minor independents and simultaneously to non-oil British major name companies and on down through a spectrum terminating with a number of nominee investment groups. The Norwegians meanwhile adopted a similar policy resulting in some remarkable prospective political and corporate alliances across the political divide. And finally, with all the enchantment that nature can muster on such occasions, oil and gas were found in many cases to be straddling the UK/Norwegian political boundary.

With the diffuse financial nature of the licensees alluded to above. the vexed question of equity finance has to be addressed. Already in many instances the ability of certain companies to meet their share of development cost has been severely limited by capital requirements, often exceeding their net worth by a factor of ten or twenty. In other words, their new found North Sea partnership liabilities might seriously threaten the financial stability of otherwise adequately financed and managed organisations.

This paradox is presently the subject of intensive financial negotiations, some apparently fruitless and others promising considerable prospects for satisfactory solutions. For example, in some instances in which the percentage ownership in the licence is substantial and the company is able, at least to some satisfactory extent, to contribute equity either directly or through subordinated borrowed funds, the oil banks have agreed to finance the bulk of the requirement on a non-recourse basis against a dedication of the oil without the benefit of technical guarantees.

In other areas where the partnership interest is minor and the financial capacity of the company to contribute even a minor proportion of its overall development costs has been inadequate for the project size, the problem has proved to be much more serious. Solutions are being sought by obtaining a guarantee from another company to support major borrowings and through government assistance in the form of guarantees. The government is hopeful that these in effect will not be financial contributions from the Treasury, but will suffice to secure the banks who will in effect contribute 100 per cent of the funds. These solutions in effect offer similar safeguards to the direct commercial technical guarantee which has been put forward by banks. For instance, in two separate successive financings in the North Sea for Occidental and the Thomson Organisation the banks were persuaded to accept a technical completion guarantee on their loan from the operator of the field, but accepted the full technical risk for repayment of their loan from the minority interest, Thomson Group.[10] As many of these solutions are especially related to the smaller independent oil companies they are discussed in greater detail in chapter 4.

The commercial loans that have been successfully structured on non-recourse bases have produced in exchange for the higher financial involvement of the banks and for the risks they are assuming, namely: (i) technical completion, (ii) a continued economic price of oil, and (iii) the validity of reserve estimates, the new phenomenon in the North Sea of the overriding royalty. Some banking commentators have gone further and described this royalty more precisely as the reward for the equity risks that banks have assumed.

Banks, and particularly oil banks, have always complained that their reward has seldom been fully commensurate with the risks they assume. Nonetheless, it must be conceded that the reward to risk ratio has been uniquely enhanced in North Sea financing by the novel introduction of the royalty and it can be further argued that, as long as the oil companies can be demonstrated to expect yields of 35 per cent or 40 per cent against highly leveraged financing, then the banks must be encouraged to provide this much needed liquidity with something better than the one or two per cent yield they have routinely accepted. As a traditional banker has put it, the normal margin is simply the rental cost of money; for risk one is entitled to charge a little more.

Other constraints on non-recourse finance
Oil lending over the years has been limited by a number of constraints that have so far precluded the banks, in the light of insufficient technical advance, from accepting certain risks. A particular instance has been in the area of tertiary recovery. It is always astonishing to realise that oil is a particularly difficult element to recover efficiently. The expected primary or natural recovery rate on proved reserves in the USA today aggregates only about 25 per cent. With oil at its present high price, greater efforts have been made to enhance that average recovery through more sophisticated systems. Secondary recovery with water or reservoir gas injection has been familiar to the banks as an acceptable system for many years and loans are now quite routinely made on properties that require artificial secondary recovery mechanisms. However, once tertiary recovery is necessary through the use of steam, carbon dioxide, fire or miscible flooding systems, the banks have tended to decline the risks. The reasons lie in the doubtful experience so far encountered with these recovery mechanisms together with the high expense usually associated with them.

A further general limitation to be observed is perhaps more financial than technical; it concerns the expense of development projects around the world, expenses which taken against the narrowing profit margins being inflicted on the oil industry, necessarily escalate the need for borrowed funds. This has produced two consequences: firstly, the companies have become more aggressive in demanding from the specialised banks greater creativity in producing limited or non-recourse loans to keep their balance sheets clear for direct borrowing not amenable to project loans; secondly, the sums they now require, traditionally financed from about 80 per cent of retained earnings, have grown so large as to place a decided burden on the banks threatening a heavy concentration of assets and risk in the petroleum sector.

Title to petroleum

Title is of vital importance in consideration by a bank of its collateral position. The legislation relating to title varies quite significantly between countries, as illustrated by the following discussion of the situation in two North American countries and the European countries.

United States of America

It is probably fair to argue that the United States, in the area of oil finance, has traditionally been less restrictive in its governing regulations than other principal oil producing nations.

Traditionally surface and sub-surface title in the United States are undifferentiated and are all vested in the owner, thereby enabling a mortgage to be publicly filed by a bank in support of its loan on those mineral rights.

If ownership of the land or minerals is in the State or Federal domain, the basic question of title does not change. In the case of Federal or State lands leased either through competitive bidding or simultaneous drawing, the leasehold obtained whether onshore or offshore, is chargeable to a bank for financing purposes. Should a foreclosure take place, a public sale is held and any party — including the bank — may acquire title to the lease at the foreclosure sale.

The mortgages or deeds of trust by the banks on federal leases are usually registered with the Bureau of Land Management. Leases on State lands are frequently filed with an appropriate State agency charged with the leasing of public lands. It is also desirable to file the interests for record in the County (or in the case of Louisiana, Parish) in which the leases are situated.

The conditions prevailing on Indian lands are different. The assignment of a lease is subject to the consent of the Secretary of the Interior and the forms of lease vary according to the Indian tribes in question. Title searches are normally made from the records of the Commissioner of Indian Affairs in Washington, D.C. and other associated official records. An assignment of a lease must be filed with an appropriate Indian agency within a prescribed period after execution of the assignment.[11]

Canada

Canada, as the result of its historical link with the United Kingdom, incorporates in its petroleum regulations a special distinguishing feature which is an important point of departure from the United States system. This lies in the question of title. With a Provincial and Federal system of government, the important question of title is difficult to summarise briefly. However taking the question broadly, title in Canada to sub-surface minerals is in the main vested in either

Provincial or Federal Crown rights. The governing regulations vary from Province to Province and where oil and gas have been discovered in quantities over time, the regulations tend to be more detailed.

In Canada, however, because Crown title cannot by definition be transferred, a quasi-transfer of the rights to explore or exploit may only be made temporarily for specific purposes with the consent of the Minister in question. In certain areas such as the Northwest and Yukon territories, the Arctic Islands and the Canadian offshore, the Federal Minister may only consent to the transfer if he deems the individual or corporation to have qualified in the first place for a licence or lease or that Canadians will have some direct benefit from the transfer.[12] In the Province of Alberta with its longer history of production and necessity for finance, assignment or transfer of title is more broadly regulated, including title transfer to allow for the pledging of oil properties as security against bank loans.[13]

Generally speaking, Canada has created the mechanics for the development oil loan through a deliberately flexible policy enabling the appropriate Ministers to consent statutorily to transfer or assignment for development finance. Canada allows most of the usual provisions enjoyed in the United States as to depletion and charge-offs, although in conditions reminiscent of the 1969 Tax Reform Act in America, Canada enacted in 1972 the 'New Act' providing for a good many reductions of former tax write-off provisions, taking the Canadian oil scene into an even tighter fiscal environment.[14]

United Kingdom
On the question of title, which at least is not in any way a modern restriction,[15] it has been legislatively determined that it will irrevocably rest with the Crown. However, unlike the statutory concessions allowed in Canada, in which as we have seen title is similarly vested, the British Government has seen fit not to modify the legislation to allow for temporary routine transfers or assignments, but has insisted, for the time being, on *ad hoc* consents which have guaranteed the security of the licence to the banks in the interests of their collateral, thereby allowing the substitution of a new licensee in the event of a default. This was somewhat troublesome in the early days of North Sea finance to those banks used to American regulations, but with repeated assurances from the British Government as to the integrity of the consents to such *ad hoc* security the problem appears for the time being to be dormant, if not completely resolved. The banks seem generally satisfied that Her Majesty's Government will, in the event of a default by a borrower, allow the assignment of the requisite interest to enable the amortisation and discharge of the loan to the banks, albeit the appointed beneficiary of the interest in the field must be acceptable to the appropriate Minister.

Norway

As with the United Kingdom, the Norwegian title question has been problematical. The Royal Decree of 8 December 1972 contains, in paragraph 49, the intention of having the field pledged as security for project loans. However, according to Norwegian authorities, the paragraph as it stands today does not in all respects provide a satisfactory basis for the pledging of the licence. Certain legal amendments are deemed to be necessary.

Project loans have not so far been tried in Norwegian waters, with all the developments so far having been unsecured direct obligations of the borrowing companies. It is principally for this reason that the whole title question in Norway has remained somewhat dormant and lacked the urgency of resolution that produced positive results in the United Kingdom.

Influence of the political climate

The influence of government policy and legislation on production financing is not a new phenomenon, as was amply illustrated in the introductory section of this chapter which dealt with the early financing of petroleum production projects in the USA.

With numerous oil companies, large and small, exploring for and discovering oil in more remote and politically volatile areas, the possibilities for the non-recourse loan have certainly been limited. Increasingly, the large looming threat of expropriation or nationalisation of a successful interest by a developing country desperately in need of a controlled source of foreign exchange is clearly a risk banks should not expect to take without viable guarantees. However, once this element is introduced, namely political risk guarantees, one has begun to stray from the notion of the true non-recourse project-related loan into areas of subsidised finance and political risk analysis.

This is a continually changing situation both in mature oil provinces, such as the USA, as well as in countries which have only recently become oil producers, such as the United Kingdom.

United States of America

It has been much more of a feature of the USA that, in the past ten years, the fiscal and legal constraints associated with the production and financing of oil have been of greater significance. The critical vehicle of change most often cited by bankers and oilmen is the Tax Reform Act of 1969 in which, firstly and traumatically, the depletion allowance of 27.5 per cent was reduced to 22 per cent and, perhaps more importantly, under Section 636 of the Act, the retained or carved-out production payment could no longer be treated as an economic interest in the property but would henceforth be treated as a mortgage loan, thereby at one stroke eliminating all the significant tax benefits accruing to it.[16] An important exception made under the Act did allow for the development production payment to retain the former tax benefits of an economic interest.

The importance of the Reform Act was twofold. Although the reserved and carved-out production payments had become increasingly a means of tax shelter, they had nonetheless served the purpose of keeping oil properties and their marketability very much in the public domain. The first structural impact of the amendment was to reduce overnight the value of the oil properties by about 20 per cent because their tax advantages had been terminated. This in turn coupled with the decline of the depletion allowance led to a severe shrinkage in both oil exploration and finance.

This rather static condition prevailed through until about 1972 when, under the impact of various international crises, the domestic demand for oil led to an upswing in the allowable production days in areas such as Texas and Louisiana, giving rise in turn to a strong resurgence of exploration and drilling activity.[17] Loan demand returned and the second golden age of development production payments appears to be on the ascendant, although one may surmise that its duration will be a good deal shorter than before, with domestic supplies increasingly more difficult to discover and exploit economically. Development production payments have today left oil finance less of a business in the USA than a specific technique for development finance. In a sense the situation has come full circle from its origins in the 1930s. The technique of the ABC carve-out had become progressively through the 1960s a project vehicle for financing acquisitions through a clever tax leverage system. Development production payments as now fiscally permitted involve oil payment from fields, both onshore and offshore, over longer periods of time. They have also been creatively modified to allow development of an undeveloped field through financial reliance on other producing fields, and since working capital is today such a vital component of an integrated oil company's financial viability, even more enterprising loan devices have been used such as the variable dedicated percentage of gross proceeds from the field as opposed to the orthodox straight line percentage dedication method.

United Kingdom

In contrast to the USA, the United Kingdom is a relative latecomer as a significant petroleum producer. As such it has the benefit of knowledge of legislation in many other countries. In judging the political issues one must necessarily keep an eye on the historical clock. Fortunately, specific governments do not last forever and short term policies, if demonstrably unacceptable, can reasonably be expected within a democratic framework to be either reversed or, at least sometimes, ameliorated. However, the proposition can be put that certain tactical political issues affecting this area have been irrevocable and will, within two distinct phases, determine for a long time the structure and strategy of exploration and finance.

The initial phase covers the period during which the first gas sector in the southern North Sea was explored and the northern licences were allocated in the two rounds of 1970 and 1971/72.[18] The notable feature of the third and fourth rounds was the praiseworthy element of democracy they contained.

Applicants and consortia were generally screened for petroleum experience, reputation, competence and the extent of British involvement. These criteria notwithstanding, it does appear now that the authorities tended to vitally underestimate and discount — but let it be said in no absolute sense — firstly, the level of financial capacity necessary in many cases to meet the future development demands and secondly, the prior involvement in the area of exploration and exploitation of petroleum of a number of companies in selected consortia.

This can now be circumstantially argued as a basic flaw in the allocation policy for licences in probably one of the most treacherous oil provinces ever encountered. It is a flaw that occurred irrespective of the government of the day. It was a flaw that even industry commentators at first discounted but progressively came to realise could become a severe handicap in development of the sector until the banks introduced innovations such as the overriding royalty discussed earlier.

The second political phase of the North Sea has unquestionably been reached under government policy announced in 1974 and broadly enacted since then. The government have contributed to the complexity by the introduction of a series of Parliamentary Acts designed to strengthen their already considerable fiscal and judicial control in the petroleum domain.

Apprehensive of the political consequences of disproportionate profits accruing to the exploiters of North Sea petroleum issuing from the fourfold increase of OPEC prices in late 1973, legislation was enacted to levy an additional tax on possible windfall profits. The so-called 45 per cent Petroleum Revenue Tax became law in 1974. This tax, on analysis, appears to have a relatively benevolent effect on smaller fields, but has an impact, predictably with progressive force, on the larger and more prolific producers.

Fortunately, from a banking point of view there were not a large number of financial packages solely reliant on oil payments at the time to be prejudiced retroactively by this legislation and its impact can now be programmed into the economic analysis of a field prior to any loan decision being taken.

There still remains the further, and administratively more complex, question of participation, namely the intention of the British Government progressively to acquire 51 per cent of all commercial fields. The policy, initially vaguely mooted as a control mechanism on a 'no loss, no gain' proposition, has steadily become accepted by the industry. The loan arrangements presently negotiated stand unaffected through state assurances by allowing the companies — and hence the banks — an undisturbed cash flow after tax, but all future loans are likely to incorporate this issue, particularly with respect to the question of royalties, which by their nature survive for the life of the field.

There are further aspects yet to be implemented under the Petroleum and Submarine Pipelines Act of 1975, namely the troublesome questions of common carrier lines and prorationing which are of deep concern to a bank considering oil payments extending out over a period of seven to ten years. If the so-called commercial overriding royalty becomes a permanent aspect of North Sea finance, the same apprehension will exist as to the true value of such a royalty discounted over an artificially protracted period of time. Future loan negotiations will no doubt produce solutions or tactical innovations to overcome these particular problems.

Conclusion

It must, in conclusion, be said that a condition to international production financing will always be a reliance upon the integrity of financial arrangements as they are affected by government policy. Clearly a tension must always exist to some extent between the autonomous authority of government to formulate policy and regulate commerce in the national interest, and the vital importance of the self interest that will determine the action of the providers of the necessary capital funds to the petroleum sector of commerce to insure that such policy and commercial contract may reasonably coincide without prejudice to either' side. It must be a pragmatic decision to accept that the stated conditions and policies, whether considered politically or economically justified, are facts the banks can, and must, deal with. Future, undefined and unquantified conditions will always be a severe hindrance in maintaining that high level of creative finance that has always marked out production finance as a highly specialised and individual sector of the international banking scene. It is clear from our area analysis that the patterns of control and regulation, although varying both in degree and impact, are increasing. *Laissez-faire* has long since been consigned to the archives of history and the concern now from a banking and finance perspective is the new problem of good security with the constant partisan ebb and flow of legislative control. Production or development finance is moving again into a new and innovative area.

References

[1] Prorationing is a technical term indicating a method of restricting oil production. It takes the form of limiting production from a well to, essentially, a percentage of its hourly full open flow rate. This need not necessarily mean an unrestricted rate as it will automatically be governed by casing and choke size. Although the restriction was calculated as a percentage it came to operate for convenience on the number of days of production within a 30 day period that would meet the overall percentage restriction. Thus one would encounter the term 'allowable days' meaning the number of days of production calculated in terms of the allocated percentage.
Open flow rates are routinely recalculated, in some cases as often as quarterly.
[2] James A. Clarke & Michael T. Halbouty, *The Last Boom* (Random House, 1972), pp 162-63.
[3] Hot oil became the designation for oil being produced outside the allowable limits and hence illegal.
[4] Returns of 300 per cent per annum were not uncommon during the early capital-hungry days.
[5] In the decade to 1940 production in Texas increased by 30 per cent, while in the following decade to 1950 it doubled to a level of over 1 billion barrels per day. See DeGolyer & MacNaughton, *Twentieth Century Petroleum Statistics,* p21 and US Bureau of Mines.
[6] Edward E. Monteith, Jnr, *Current Status of Mergers and Acquisitions in the Petroleum Industry,* Paper SPE 2069, (American Institute of Mining, Metallurgical and Petroleum Engineers Inc, 1968).
[7] For a full technical description of all the basic components of petroleum reservoirs, a good treatment can be found in Norman J. Clarke, *Elements of Petroleum Reservoirs,* Henry L. Docherty Series (Privately published for the American Institute of Mining Engineers, 1969).
[8] Willis G, Meyer, *Use of Oil and Gas Property Appraisal Reports,* Special Paper No. 1 (Willis G. Meyer & Associates, Dallas, Texas).
[9] An analysis of the issues specifically related to the North Sea is given in N. A. White, 'The Challenge of Non-Recourse Finance', *Journal of The Institute of Bankers,* February 1976.
[10] This financing refers to the Piper field in which Occidental has a 36.5 per cent interest and is also the operator. The other members of the consortium are Thomson 20 per cent, Allied Chemical 20 per cent, Getty Oil 23.5 per cent.
[11] Alfred E. McLane, *Oil and Gas Leasing on Indian Lands* (F H Gower, Denver, Colorado, 1955).
[12] Bank of Montreal Oil and Gas Department, *A Guide to Oil and Gas Operators in Canada,* revised edn. (July 1974), p6.
[13] *Ibid.* p16. Permitted under section 82 of the Bank Act of Canada.
[14] *Ibid.* Comparative appendix, final unnumbered page.
[15] Petroleum (Production) Act 1934, section 1(1).
[16] Granville Dutton, *The Effects of the 1969 Tax Reform Act on Petroleum Values,* Manuscript SPE 3261, (American Society of Petroleum Engineers, 15 October 1970).
[17] 1976 allowables in Texas and Louisiana are both 100 per cent of production.
[18] There were, in fact, four official licence allocations in 1964, 1965, 1970, 1971 and 1972 (when the second part of the 1971 round was completed), and the fifth round announced in 1976.

6

Tankers and Offshore Drilling Rigs

Long before the modern oil industry came into being animal, vegetable, and some mineral oils were shipped around the world. They were transported, like most other cargoes, in specially made containers, such as wooden barrels. The first cargo of oil to cross the Atlantic travelled in the little 224-ton brig *Elizabeth Watts* in 1861. Demand in Europe grew rapidly so that, by 1864, 32 million gallons were being shipped across the Atlantic in these primitive containers.

Historical development of tankers

Soon the barrels were replaced by larger and larger iron tanks which were fitted eventually into the hull of the ship. Probably the first vessel that actually carried oil across the Atlantic in bulk using tanks fitted into the holds was the 794-ton sailing ship *Charles*, which traded successfully between 1869 and 1872. She was equipped with fifty-nine iron tanks each holding about 13 tons.

The first modern oil tanker was probably the S.S. *Gluckauf* built on the Tyne in 1886, gross tonnage 2,307 tons, 300 feet long, and driven by triple-expansion engines and coal-fired boilers. The many advantages of such vessels were quickly recognised. By 1891 there were nearly eighty true tankers plying between American, Russian and European ports.

At the beginning of the century, the world tanker fleet had grown to half a million tons with an average deadweight of 5,000 tons per tanker; by 1914 the total tonnage had increased to 2 million tons deadweight, and the average size to something over 6,000 tons per tanker. By 1939 the world tanker fleet had grown to more than 1,500 vessels totalling 16.5 million tons, and the standard tanker was 12,000 dwt with a speed of eleven knots.

The vital need for oil during the Second World War gave a tremendous impetus to tanker building. The lead was taken by the USA which, having developed a standard tanker of 16,500 tons with a speed of 14.6 knots, proceeded to produce such vessels in very large numbers. Between 1942 and 1945 nearly five hundred of these ships, known as T2s, were built. Thus, despite heavy losses, the world tanker fleet had risen to 24 million tons by the end of the Second World War.

After the war a number of influences combined to create a revolution in international oil transportation. The growth of the Middle East as a producing centre, the practice of building refineries in the consuming areas instead of near the oil fields, and the enormously increased demand for oil in the industrialised regions led to a five-fold increase in tanker carrying capacity between 1950 and 1970, as shown in table 14.

Table 14: Development of world tanker fleet — total tonnage

expressed in millions dwt

1950	23.4
1955	37.4
1962	63.3
1965	80.2
1970	131.5
1974	215.6
1975	254.3
1976	290.9
1977	321.0

Perhaps of greater economic significance than the growth in tonnage has been the dramatic increase in the size of individual ships. Data relating to the progressive growth in the dimensions of tankers are shown in table 15.

Table 15: Development in size of tankers

Dimensions expressed in feet

Year	Dwt	Length	Beam	Draught	Freeboard
1945	16,500 (T2)	523	68	30	9
1950	28,000	624.	80	33	12
1960	70,000	820	112	45	13
1965	120,000	886	139	52	15
1968	200,000	1,075	155	62	18
1970	332,092	1,133	175	81	24
1975	484,332	1,243	204	92	26

Since the 1930s there has been a significant change in the ownership of tankers. Initially oil companies built their own ships, but progressively tankers have been taken over or built by non-oil companies or individual shipowners, who now control over 50 per cent of the world's tanker fleet.

Financing of independently owned tankers

During the 1930s Greek and Norwegian shipowners, in particular, encouraged the trend towards independent tanker fleets. In the last 15 to 20 years the big independent tanker owners have come into prominence for two main reasons. The shipowners saw opportunities for establishing a continuing and long term transportation relationship with the major oil companies. The major oil companies recognised the advantages of having a pool of tankers owned by the independents to service up to, say, about 50 per cent of their total shipment demands at any given time. One advantage was the so-called off-balance sheet finance. By allowing or encouraging the independents to build up their fleets, either by contracting for their own vessels or buying contracts or ships from the majors, the oil companies freed capital resources for use in exploration, production, refining, petrochemical industries and distribution. In future the oil companies are likely to use company fleets for less than 50 per cent of their transportation requirements. The balance will be met by the fleets of independent tanker owners, oil producing countries and, probably to a lesser extent, by ships belonging to developing nations. Some of these will be in the form of joint ventures between oil producing countries and independent tanker owners.

The independent owners have been able to raise the necessary finance to contract for new ships or to buy existing ships on the strength of medium to long term charter-parties from major oil companies whose ability to meet charter payment obligations was undoubted by the banks.

How has the expansion of the independent tanker owners' fleet been financed? There are basically three sources of finance available to independent tanker owners:

— owners/shareholders equity;
— shipyard finance or government and government sponsored credits;
— commercial bank lending.

As a general guideline the banks usually advanced up to 50-60 per cent of the cost, whereas government sponsored schemes covered 70-80 per cent of the cost of each vessel. This left a shortfall of 40-50 per cent in the case of bank lending and 20-30 per cent in the case of government sponsored finance. This difference had to be put up by the owner as his equity or risk capital in the venture, although sometimes the shipyard concerned would contribute towards it. The larger the equity portion the less the risk for the main lender in case of a drop in value or forced sale of the vessel.

Equity finance

There are various ways in which the equity portion can be raised. Scandinavian owners often formed partnerships among their local associates and were able in this way to raise the required equity. An established owner would provide the funds from his accumulated cash resources or from the sale of other ships and other investments. Sometimes the shipyard's bankers or other lenders are prepared to grant loans against a second mortgage on the vessel. The first mortgage lenders may refuse to allow the owner to do this if, in their opinion, the shipowner's own stake becomes too small. A high gearing increases the risk of the whole project because higher income is needed to service the increased repayments and interest burden. No

Table 16: Government assisted finance for foreign shipowners
(for large vessels over approx 6,000 dwt)

	Amount	Term	Interest	Security	Currency
Denmark[†]	70%	7 years	8%	1st Mort.[1]	D.Kr.
Finland[†]	70%	7 years	8%	1st Mort.	US$ F.Mk.
France[†]	70%	7 years	8%	1st Mort.[2]	F.F.
Italy[†]	50%	15 years	$5\frac{1}{2}$%	1st Mort.	Lire
Japan[†]	70%	6 years	$8\frac{1}{2}$%[4] approx.	1st Mort.	Yen
Netherlands[†]	70%	7 years	10%	1st Mort.[3]	D.Fl.
Norway[†]	20%	7 years	$8\frac{1}{2}$%	2nd Mort.	N.Kr.
Poland	*	*	*	*	*
South Korea	*	*	*	*	*
Sweden[†]	70%	7 years	8%	1st Mort.[1]	S.Kr/$
United Kingdom[†]	70%	7 years	$7\frac{1}{2}$/8%	1st Mort.	£
United States of America[†]	30%	7 years	7/$8\frac{1}{2}$%	Negative pledge in respect of vessels	US$
West Germany[†]	70%	7 years	10%[4] approx.	1st Mort.	DM
Yugoslavia	to 50%	10 years	*	1st Mort.	Yu.Din

†Signatory to OECD agreement.
*Subject to negotiation.
[1]1st Mortgage on 50%; other security for remaining 20%.

hard and fast rules exist as each case has to be assessed on its own merit. For traditional shipping companies, sale of written-down old or uneconomic vessels is the recognised method for raising the cash needed for the equity in a new ship.

Government and government sponsored facilities

Public and semi-public agencies have become a major source of finance.[1] It is important, especially in the 1970s, to understand the background to this development. With the increasing demand for tankers during the last twenty years, traditional shipbuilding nations were eager to capture as much of the market as possible. They met stiff competition from those developing countries which had decided

Source	Other
Danish Ship Credit Fund	Post-delivery only. Guarantee required from Export Credit Council.
Finnish Export Credit	Currency risk is with Export Credit.
Commercial Banks/COFACE	Post-delivery. Pre-delivery through yard's banking pool.
Instituto Mobiliare Italiano	
EXIMBANK (55%) Japanese City Banks (45%)	Average amount currently financed is 60%. Funds lent through shipyards.
Finance from yard/commercial banks. Interest subsidy paid 5/6 weeks in arrear to yard and thence to shipowner.	
A/S Laneinstitutet for Skipsbyggeriene	Post-delivery only.
Arranged through Centromor Finance from banks	No specific scheme. Credit facilities evaluated on an individual basis. Certain vessels 'cash only'.
*	Scheme currently under consideration.
Shipyards	
Commercial Banks/ECGD	
Ex-Im Bank	No single financing scheme. Minimum contract price US$3 m. Each case is considered on its merits. 10% cash payment in loan participation programme is required. Remaining 60% comes from private sources.
Commercial Banks/Ship Mortgage	
Export Credit Fund.	

[2] Additional collateral securities may be required for an amount equal or up to 30% of the loan.
[3] 1st Mortgage on 60%; other security for remaining 10%.
[4] Weighted average of funds from different sources.

that the construction of tankers was a natural and relatively simple way into the industrial world. Erection of shipyards with auxiliary supporting industries looked a good and rapid way to provide employment for large numbers of people, and could form the basis for further technological developments. In order to attract orders from the shipowners the shipyards focused more and more on credit facilities as a sales tool. As credit terms were made increasingly attractive to prospective buyers, normal banking facilities became inadequate and shipyards had to turn to their governments for assistance. Governments were anxious to protect the employment of thousands of workers, especially in districts with few alternative employment possibilities. They were also keenly interested in securing valuable foreign currency earnings from export contracts. Through direct loans or financial guarantees many governments enabled their shipyards to offer attractive package deals.

Government subsidies and credit arrangements vary from country to country, and some of the terms and conditions available in the spring of 1977 are outlined in table 16. In principle, most countries adhere to the present OECD rules. Under these countries may grant loans for a period of up to seven years. The loans must be repaid in equal instalments at regular intervals, normally of six months but with a maximum of twelve months. The shipowner must find equity of 30 per cent of the cost price, and the interest rate on the loan must not exceed 8 per cent including bank charges. However, as the order books run out there are many examples of special schemes introduced by governments in shipbuilding countries. Under present OECD rules member countries are allowed to offer any type of credit facility for new building contracts for domestic owners. Export contracts for developing countries are also excluded provided there is an element of aid of not less than 25 per cent of the contract price.

In the 1960s and up to the end of 1973 many governments and credit agencies as well as banks frequently paid little attention to the credit-standing of the buyer. In retrospect it is easy to see how faith in continuing inflation and ever-increasing costs and values in money-terms tempted shipowners to enter into new building contracts without sufficient equity and without simultaneous freight contracts with first class chartering companies. Equally, yards were anxious to obtain contracts with or without simultaneous credit and charter agreements. The international crisis of 1974/6 led to massive cancellations of new building contracts for oil tankers. Some shipping companies foundered, others were kept alive by shipyards, governments and banks agreeing special arrangements, moratoria and joint ventures.

Banks' attitude to risks

It is important to realise that the traditional role of commercial banks has been to provide short term facilities to industry and commerce. Only during the past decade or so have such banks become used to setting aside a proportion of their resources for medium term commitments for industry and shipping. Flourishing world trade, inflation, and increasingly tough competition between banks with easy access to the eurocurrency markets led to a rapid increase in medium term loans from the late 1960s.

The following paragraphs will describe the main considerations and criteria used by commercial banks when considering loans for tankers and offshore drilling rigs.

Financial standing of borrower

Tanker shipping is a cyclical business, subject to world supply and demand. It is of paramount importance to be aware of the general and financial standing of the borrower at all times. In most cases there exists a long and good relationship between the client and the bank which greatly assists decision making. However, independent shipowners have often been reluctant to disclose their position, plans, and commitments fully, and it is vital for a bank to keep constant check to ensure that the individual owner or management group of the company is of a high calibre. Banks often ask for cash flow projections from shipowners, usually to cover the period of the proposed loan, and charter-parties to first class charterers will provide evidence of future cash flow. The bank can then evaluate the financial standing of the borrower over the proposed loan period. Where the shipowner owns vessels which do not have long term charter-parties the cash flow projections will be less certain, and a bank will undertake a sensitivity analysis to ascertain a shipowner's financial position in case the market should vary from the owner's projections. The opinions of charterers, agents and other users of the companies' services are also important.

The formal ownership of tankers can take several forms. The individual shipowner, for example, may own the ships in his own name, thereby assuming full liability for any borrowings. He may form a partnership, create a limited liability company together with other investors, or invite the general public to subscribe capital to a new publicly quoted company where the initiator, or shipowner, assumes the role of a manager with little or no personal financial responsibilities.

One ship company

If the owner owns and operates his fleet by means of a one ship offshore company without firm responsibility for borrowing being accepted by himself or the main asset-holding company, the bank will face a greater risk than it would from lending to an established ongoing concern.

One ship companies are common among Greek, American and Far Eastern owners, who, for fiscal and other reasons, establish Panamanian or Liberian companies with nominal capital for the purpose of owning one vessel. The vessel is then usually bareboat-chartered from the one ship company to the shipowner/operator on a long term basis. He will then time-charter out the same ship for various periods and at various rates to suit his particular expectations or cash flow. Many of the Hong Kong owners entered into such *Shiqumisen* deals with Japanese owners, and were able to raise finance by assigning the bareboat charter to the lending bank, in addition to giving the bank a first mortgage on the vessel. Companies such as the Sanko Steamship Companies and World-Wide Shipping (Mr Y. K. Pao) of Hong Kong have made extensive use of this method.

Country of registration

The lending bank must also take into account the flag under which the offered ship is registered. The laws under which a mortgagee can foreclose or take possession of the ship in the event of default vary from country to country. Most established maritime nations have laws enabling effective legal action, whereas the legal procedure in other countries can drag on for years. If a lender has to arrest a ship, it is very important that the port where the arrest takes place lies in a country with an effective and reliable maritime law.

Size and nature of oil tankers

Oil tankers tend to be classified into the following categories for the purpose of bank loans:

    Crude carriers of the following sizes:

        (a) up to 100,000 dwt
        (b) from 100,000 to 200,000 dwt
        (c) from 200,00 to 350,000 dwt (VLCC)
        (d) over 350,000 dwt (ULCC)

    Product carriers to transport refined products.

Basically, banks take the same approach to requests for tanker finance in both main categories, but the question of size (outlined in a-d) may also bear on their judgment. In considering these categories world supply and demand must be assessed. If in future large refineries and petrochemical plants are to be built in close proximity to the oil fields, the demand for tankers, particularly for the large independently owned variety, will diminish, with a corresponding increase in demand for product carriers, parcel tankers and chemical carriers.

LNG — Liquefied natural gas carriers
These are highly specialised and technically advanced ships constructed for only one trade, the carrying of LNG from production centres to consumption centres. At a cost of, say, $140 million for a 125,000 m³ carrier, or perhaps three times the price of a ULCC, it is especially important that such ships have continuous employment. All but the very strongest companies would have difficulty in obtaining long term finance without a firm, reliable, long term charter-party with a first class charterer. The LNG carrier depends entirely on the completion of all installations in both the producing and the receiving country. (The risks associated with the complete LNG System are discussed in chapter 8.) Above all, the shipping company must be able to provide highly qualified technical personnel to operate and maintain the vessels. The charter-party should ideally be 'watertight' to ensure depreciation of the vessel over, say, fifteen to twenty years, with few, if any, escape clauses for technical or other reasons.

This type of finance does not readily suit the commercial banking system. At the time of writing three 125,000 m³ carriers have been financed through the cooperation of international commercial banks, the shipyard and a government sponsored agency. Given the right structure and tight charter-party terms many LNG carriers will probably be financed by long term finance institutions like insurance companies and pension funds.

Drilling rigs and drilling ships
Since the discoveries of offshore oil deposits various types of drilling equipment have been used to explore and extract the oil. The most common vehicle is the jack-up rig which normally is fixed in one particular shallow-water area. As the exploration moved to deeper waters in rougher areas, for example the North Sea, the need for stronger and better equipped drilling rigs became apparent. This resulted in the development of the semi-submersible drilling rig and the dynamically positioned drilling ship. Both are expensive units and require highly professional management and crews. Until 1975 charters of two to five years were customary, with high charter rates, but, since then, shorter periods have had to be accepted due to fall in demand.

The British and Norwegian shipowners were in the forefront of these developments and were among the first to place orders for rigs and ships. Their proximity to the North Sea was no doubt a contributing factor, and they regarded the development of rigs as a natural extension of their shipping activities. Consequently, they approached their financing requirements in much the same way as they had tackled tanker finance in the past. The international banking community, having had little or no previous experience in financing rigs, initially agreed the tanker finance concept and several syndicated rig financing deals were made in 1972/74 of up to 80 per cent of cost price, sometimes without any substantial company having the ultimate responsibility or the financial strength to meet the monetary burden if anything should go wrong.

Risk assessment for rig projects is conducted by the banks and by the government agencies in much the same way as for tankers. Again, the emphasis can be different according to financial strength, employment or shipyard work, the need for orders or for foreign currency earnings. Because of the difficulties of estimating the real demand for oil rigs and drill ships, the delays for technical reasons of North Sea programmes, the reduced demand for oil, and the unexpected restrictions imposed by the British and Norwegian Governments on exploration and production, it became apparent in late 1974/early 1975 that too many rigs had been ordered. As a result of the downturn in the tanker market and consequent problems facing owners and banks, it became almost impossible to raise banking finance for rigs without firm employment and proven management. The situation became very serious in a number of cases, requiring major re-negotiations and cooperation between owners and financial institutions.

Characteristics of loan finance

As commercial bank loans are invariably more expensive than the subsidised finance from government agencies described earlier, the latter method should be investigated fully before commercial loans are considered.

Amount of loan
The loan amount is usually expressed in, and limited to, a certain percentage of the cost price or market value of the tanker at the time of delivery, or at the time an existing tanker is purchased. The advance against each ship has varied throughout the years, from a small percentage up to 80-100 per cent in extreme cases, although in the last two decades the usual amount advanced has been limited to 50-60 per cent of cost/value on newbuildings and 40-60 per cent for secondhand tankers. To advance up to 80 or 100 per cent is regarded by most bankers as wrong in principle — the owner should put in a reasonable amount of equity — and such advances leave no risk cushion for the lending bank. Should the ship value fall — and

this can easily happen — the bank is overexposed at once and must make provisions. This means that the borrower has no risk himself unless he has given his personal guarantee which, in theory, should give some additional security, but, in practice, means very little.

Loan period

The duration of the loan period has changed over the years. Decisions on this question are always taken in the light of conditions at the particular time in the financial markets, in world trade generally and in oil markets in particular. However, certain patterns have been established and developed over the years. Generally speaking, commercial banks expect a reasonable, quick return from their advances, but accept that for high costs units often linked to a term charter-party, a certain flexibility must be given.

Since 1945 we have seen three distinct patterns and one shortlived variation:

(1) A loan period of 5 years; i.e. the loan is repaid by equal half-yearly instalments of one tenth of the original loan amount, first instalment due six months after the loan is paid out.

(2) A loan period of 5 years, but instalments are repayable as if the loan were granted for 8 years; i.e. each equal half-yearly instalment represents one sixteenth of the original loan amount with a final payment of three eighths of the original loan amount 5 years after pay-out. In many cases there is an understanding between the borrower and the bank that the 'balloon' outstanding after 5 years could be refinanced over 3 to 5 years. The decision for an extension of the loan would be taken by the bank in the light of market conditions at the time, assessed against the current standing of the borrower, condition and market value of the ship, and employment possibilities.

(3) A loan period of 10 years; i.e. the loan is repaid by equal half-yearly instalments of one twentieth of the original loan amount.

(4) For a short time during the years 1972/74 a number of banks were willing to increase the loan period up to 12-14 years, sometimes with a grace period of 2 years without repayment after drawdown of the loan. Another variation was to allow the borrower to postpone any two half-yearly instalments of his own chosing to the end of the loan period, thus granting the borrower a certain flexibility should the cash flow come under pressure.

There are certain merits in granting loans under the alternatives stated above. Alternative (1) places a great restraint on the borrower, bearing in mind that according to tax rules and commercial practice tankers are depreciated over 12-20 years, whereas the useful working life of a tanker has for a long time been estimated at around 25 years.

With the building of bigger and more costly units in the early to mid 1960s, alternative (1) was replaced by the more favourable alternative (2), which went some way to assist shipowners in placing new orders. Both alternatives have the advantage that unless the borrower had the necessary strength to finance the difference himself, or felt he could manage with such stringent terms, the deal would not be entered into.

Currency of loan

The normal procedure for banks is to advance loans in their own currency. For various reasons, particularly after 1945 when each country's authorities imposed or continued to operate severe restrictions on foreign currency transactions, the banks were limited to financing tankers built in their own countries. The exceptions were the American banks, who were free to lend almost anywhere in the world until 1963 when Interest Equalization Tax (IET) was introduced, virtually putting a stop to their financing of foreign shipping companies from the USA. This change in the USA triggered the rapid growth of the so-called eurodollar market with the City of London as the principal centre. More correctly this market should be called the eurocurrency market since several other international currencies are involved as well as the dollar. The eurodollar market — further details of which are given in chapter 13 — has been steadily growing for the last 10-12 years and has successfully withstood all pressures and crises. The growth of the eurodollar market was greatly assisted by the rapid increase of new banks from the United States, Japan and the European countries, which were being set up in London during the period, together with the expansion of the existing offices in London of the major American banks. This growing body of banks in London made advances in eurodollars to non-American borrowers. In theory this type of lending involves additional risks for the banks involved compared with the normal risk present when a bank lends domestically in its home country. There is a possibility that a major crisis could arise causing the market to be closed for certain periods, so that banks might find themselves in a situation where they cannot borrow to finance loans already committed. The problem is complicated by the fact that it is an international market and no one government is responsible for the control of the market, or will provide a 'lender of last resort' in the form of a central bank. In practice the market has operated flexibly and reliably throughout periods of considerable market instability and has become a major source of international finance. The use of non-domestic currency also imposes risks on the borrower, and the lending banks must carefully assess the risk the borrower undertakes in this respect.

Until 1967 and in some cases later the majority of charter-parties were expressed in sterling. Some sterling charter-parties remain in existence today, usually being long term arrangements which have not yet run out. One can readily understand the risks involved if an owner signs a purchase contract for a tanker in one currency, arranges finance in another and receives the freight income in a third currency. These problems of exchange risk are made worse if the income is fixed on a long term charter-party denominated in a weak currency, such as has been the recent experience of sterling.

Rates of interest
The rates of interest charged by the banks over the years have been continuously adjusted in line with general market changes. In arriving at the rate to be charged or a margin over a basic lending rate, the bank normally considers factors such as length of loan, risk associated with loan, an adequate return on capital and competition for the business.

There are two principal types of loan methods in use: a fixed rate for the duration of the loan or a floating rate for the duration of the loan, that is the margin agreed at the outset and is added to a base rate, which is determined by money market rates of interest. Such a base rate may be determined from interbank money market lending rates, central bank discount rates, or determined on some other formula. In the eurocurrency market interest is normally charged under a floating rate method. It is normal to fix the rate on the loan for, say, six months at a time. This six month period is called the rollover period and is usually set to coincide with the payment of instalments on the loan. At the end of each rollover period of six months — the rollover date — a new base rate is set for the next period. The usual basis of setting the rate is to refer to the London Interbank Offered Rate for US dollars, conventionally shortened to the term 'LIBOR rate'. Once this rate has been set for the six month period, it is not changed until the next rollover date, even though the market rate for six month dollars will be changing each day during the rollover period.

Commissions, fees and other costs
In addition to the interest charges, a borrower will usually have to pay certain commissions, fees and costs. These are:
  —a negotiation commission which is levied when the bank enters into the financing commitment with the borrower, sometimes one or two years before pay-out;
  —a commitment commission which is charged at intervals of three or six months on the undrawn proportion of the loan

from date of commitment until pay-out (drawdown) of the
loan;
—fees and costs; i.e. lawyers' fees, travel costs, printing of loan
documents, etc. These are as a rule carried by the borrower.

Security for loan finance

There are two basic types of loan — unsecured and secured. To lend to
the owner or owning company for the purpose of part-financing a
tanker without specific security in the ship — corporate lending
— should only be made if the lending bank feels it has an intimate
and thorough knowledge of the borrower, his overall position and
track record. This type of loan can involve a high risk and is not
generally acceptable to banks.

Secured loans are by far the most common form of ship-financing.
Types of security acceptable to banks include:
  —a first preferred mortgage on the ship;
  —a second preferred mortgage on the ship;
  —unconditional guarantees from a third party, i.e.
    a state
    a semi-state agency
    a first class bank
    a first class insurance company
    a first-class corporation

Although mortgages are registered on the ships the loan is by no
means 'fully secured'. It is important to remember that values of the
mortgaged assets fluctuate almost continuously, at times violently.
Assuming that the advance against the ship is made with a conservative, generally acceptable margin — say within 50-60 per cent of cost
or market value at the time of pay-out — the bank should be
reasonably protected.

Mortgage on the ship or rig

The country of registration of the ship or rig is a key consideration
in terms of the value of the mortgage as security, as discussed in an
earlier section. It is essential for a lender to engage first class
international lawyers to arrange the formal registration of mortgages,
especially if it becomes necessary to instigate foreclosure proceedings.

Since the mortgaged ship constitutes the main security for the
loan, the protection of the security is important. The lender therefore
normally requires, and the shipowner generally agrees, that the
following insurances are taken out and assigned to the lenders.

Hull insurance: This policy covers the hull, machinery and complete
  equipment. In addition it specifically covers loss of, or damage to,

hull or machinery directly caused by the following:

Accidents in loading, discharging of cargo or fuel;

Explosions on shipboard or elsewhere;

Bursting of boilers, breakage of shafts or any latent defect in the machinery or hull;

Contact with aircraft;

Negligence of master, officers, crew or pilots, provided such loss or damage has not resulted from want of due diligence by the assured, owners or managers.

Hull interest insurance: By hull interest is meant interest which the owners have in their capacity as owners of the insured ship, and which is neither covered by hull insurance nor relates to a particular charter-party. This insurance is limited to a maximum of 25 per cent of the hull insurance.

War risk insurance: This insurance covers the risks excluded from the hull insurance, and loss or damage to the ship caused by

— hostilities, warlike operations, civil war, revolution, rebellion, insurrection or civil strife arising therefrom;

— mines, torpedoes, bombs or other engines of war;

— strikers, locked-out workmen or persons taking part in labour disturbances, riots or civil commotions;

— persons acting maliciously.

Third party liability insurance (P & I — Protection and Indemnity) covers liability and other loss, provided the loss has occurred in direct connection with the running of the ship, irrespective of whether it is caused by marine or war perils.

The rule for all of the above types of insurance is that where the interest to which the insurance pertains is mortgaged, the insurance also protects the interest of the mortgagee, provided, however, that the mortgagee's rights against the insurer shall not exceed the rights of the owner.

Insurance of freight etc. — There is a variety of ways in which the owner may insure against loss of voyage freight, loss of time-charter hire, loss caused by strikes, extraordinary costs, and freight interest. These matters are discussed further in chapter 10.

Assignment of charter-parties

Where charter-parties exist, the bank requires these to be assigned to ensure that the cash flow emanating from the charter-party shall be channelled through the bank to be used, in whole or in part, to service the outstanding loan.

There are basically three forms of employment contracts: bareboat or time charter; single or consecutive voyage charter; contract of affreightment.

Bareboat charter — The owner is putting the ship at the charterer's disposal for a specific period of time. Charterers obtain complete control of the vessel which they are operating as if she belonged to their own fleet. Freight is payable for the entire charter period regardless of 'off-hire'. The vessel shall, at the expiration of the charter period, be re-delivered to the owners in the same good order and condition as when delivered, ordinary wear and tear excepted.

Time charter — The shipowner provides and offers the full services of a named vessel for an agreed period. The owners shall exercise due diligence to maintain the vessel in a seaworthy condition as well as to keep her in a thoroughly efficient state as regards hull, machinery and equipment during the period of the charter-party. Time-charter hire runs continuously unless the vessel is 'off-hire' as a result of extraordinary circumstances, or periodic drydocking.

Voyage charter — Under this type of charter the shipowner undertakes to put a named vessel at the charterer's disposal for the carriage of a full cargo or part cargo from one or more ports to named port(s) of destination or ports within a certain range.

Contract of affreightment — The shipowner undertakes to move, within an agreed period, a specified quantity of a commodity between ports. The contract usually contains a quantum-range for each shipment but otherwise the entire planning, programming, and execution of the transportation is taken care of by the ship-owner, who also carries all risks related to the transportation. The contract is normally not related to a specific vessel, and the owner is allowed to provide the service of different vessels. This contract form normally requires considerable flexibility on the part of the shippers, carriers and receivers, and also from the charterers as far as supply of cargo to be shipped and the receiving facilities

Table 17: Characteristics of employment contracts

	Bareboat	Time	Voyage	Contract of affreightment
Specific vessel	Yes	Yes	Yes	No
Freight payable on	Summer dwt	Summer dwt	Actual cargo carried	Actual cargo carried
Officers appointed by	Charterer	Owner	Owner	Owner
Crews wages paid by	Charterer	Owner	Owner	Owner
Fuel paid by	Charterer	Charterer	Owner	Owner
Port charges paid by	Charterer	Charterer	Owner unless otherwise agreed	Owner
Insurance paid by	Charterer	Owner	Owner	Owner
Maintenance of vessel	Charterer	Owner	Owner	Owner

are concerned. It is, therefore, generally preferred for very large shipments of more simple commodities such as crude oil, coal, iron ore etc. where substantial buffers can generally be found and are easily available both in the loading and discharge ports.

Table 17 summarises the main characteristics of these various contracts.

Documentation for loan

This forms an integral part of the security for the loan. Unless possible eventualities are adequately covered the security is significantly reduced.

The necessary formal documentation related to the financing is basically the same when the vessel is financed through either of the two main sources mentioned above and can be listed as follows:

    Loan agreement
    Mortgage deed, and, when necessary, deed of covenant
    Classification certificate
    Certificate of entry in the ship register
    Insurance policies for assignment
    Charter-party for assignment
    Necessary licences from owner's own authorities, if necessary, for exchange control and foreign borrowing purposes.

Current problems and the future

The tanker industry was naturally one of the first industries to be hit — and hit severely — by the effects of the sudden, massive increase in the price of oil towards the end of 1973. Freight rates for spot and voyage charters fell dramatically and the vast extent of the tonnage surplus has prevented any significant recovery in the market despite a strong resurgence in oil liftings during 1976. The poor employment prospects for tankers in turn depressed the secondhand market, and the realisable value of some modern tankers fell by as much as 75 per cent. The security represented by a tanker mortgage was therefore much undermined, and as more and more vessels came off profitable period charters and entered the spot market the liquidity problems of many independent owners became acute. In a number of cases these owners had also invested in offshore drilling rigs, and found the market for these expensive vessels equally depressed. Clearly, with no expectation of a short term recovery in these market, the outlook for a number of independent owners is grim, and commercial banks and other financing institutions are having to deal with a continuing series of problem loans.

Unquestionably we are still in a depression of unusual severity, but the exposure of the commercial banking industry to the most vulnerable sector of shipping should not be exaggerated. In the first place over half of the mortgage debt outstanding on tankers and offshore drilling rigs is government guaranteed or insured. The remainder of the debt is, moreover, spread throughout a very wide range of banks, reflecting the international nature of the shipping industry. In addition, it was a common practice in 1972/73 to syndicate loans for large tankers among ten, twenty or even more banks, thereby further diluting the risk to individual institutions. Because the exposure of commercial banks has not been dangerously large, the problems which have arisen in connection with shipping loans have by and large been handled and contained through holding operations.

In Norway, where the shipping community has been highly vulnerable to the depressed market for tankers and drilling rigs, the prevention of a spate of sales of ships at very depressed prices has been due in large part to the establishment in December 1975 of the Guarantee Institute for Ships and Drilling Rigs. The Institute is owned 60 per cent by the Norwegian Government, 20 per cent by Norwegian shipowners and 20 per cent by Norwegian financing institutions, and through the provision of loan guarantees it is able to encourage holding operations for fundamentally sound and well-managed shipping companies with modern tonnage which would otherwise have to be sold cheaply abroad to provide liquidity. Since over 80 per cent of the Norwegian ocean-going merchant fleet has been financed outside Norway, the Institute has become involved in major negotiations with foreign shipyards and creditors, leading in a number of cases to arrangements for the restructuring of companies and loan moratoria.

Another important organisation which has grown out of the tanker crisis is the International Maritime Industry Forum (IMIF), which was set up at the beginning of 1976 by a group of oil companies, independent shipowners, shipbuilders and banks to discuss ways of alleviating the crisis. The IMIF has, in particular, done much to encourage the cancellation of tanker newbuilding contracts and to inform governments of the nature of the problems being faced in shipping and shipbuilding.

The events of the past two or three years have, not surprisingly, caused financing institutions to adopt a cautious attitude to new tanker and rig finance propositions. With recovery in the tanker and rig markets still some way off, commercial banks are likely to pay great attention to borrowers' financial standing, cash flow and overall commitments. They will wish to see firm employment for tankers, producing an adequate level of cash flow, or alternatively clear evidence that the borrower's overall cash flow is sufficient. Banks will in many cases require additional security and will want wider interest margins, and the documentation for loans shows signs of becoming increasingly restrictive.

The major worry now for both shipowners and their bankers is the vast over-capacity of the world shipbuilding industry. Most shipyards are rapidly running out of orders, and there is strong political and social pressure on the governments of shipbuilding countries to encourage their continued employment through the provision of ever more generous subsidies and credit terms. At the same time new shipbuilding countries like South Korea, Brazil and Indonesia — as well as the Republic of China and the COMECON states — are increasing their yard capacity. Unless there is international agreement at industry and government level on a phased and coordinated reduction of capacity, the threat will continue of the construction of more heavily subsidised surplus tonnage. The removal of this threat would do much to restore the confidence of financing institutions in the prospects of international shipping. Inevitably, the shipping industry faces a period of restructuring, but this should enable a stronger independent tanker-owning sector to emerge in the 1980s. Commercial banks and other financing institutions will continue to be willing to provide finance on a sound basis for this industry, as long as they are sure that the shipbuilding industry is not being maintained at an artificially high level and that new ships will only be built in response to commercial demand.

References

[1]Chapter 11 discusses the general subject of export finance and associated credit insurance schemes available in the industrialised countries.

7

Pipelines and Processing Plants

Pipelines in the petroleum industry may be used to transport crude oil, natural gas liquids or natural gas from the oil or gas fields or tanker terminals to the consumer or the processing plant; alternatively they may be used to distribute refined products to the consumer or to further processing plants. The cost may vary from only a few million dollars, for a line joining a refinery and a neighbouring petrochemical facility, at one extreme to several billion dollars, for projects such as the major Alaskan oil and gas lines, at the other.

Processing plants, in the context of the industry, are principally refineries which break down crude oil into its constituent butanes, propanes, gasolines, kerosines, gas oils, fuel oils, and perhaps bitumen, sulphur, petroleum wax etc. The term also includes, however, petrochemical plants which receive part of the output of a refinery, perhaps gases, naphtha or part of the kerosine/gas oil production, or which may run on natural gas or natural gas liquids. Such plants may produce an enormous range of hydrocarbon-based petrochemicals. Here, too, the cost may range from a few million dollars, for an extension to an existing plant, to several billion dollars, for a large refinery and associated petrochemical plant built on a virgin site.

Historically, major projects such as these were largely financed in equity form. Since the war corporate balance sheets have become even more highly geared, and individual projects have been financed more and more in debt form. In recent years, highly specialised forms of debt have been developed, which provide for repayment only out of the funds generated by the project, and not out of funds arising from the remainder of the borrower's business.

General approach to financing

The financing of pipelines and processing plants poses two fundamentally similar problems. In the first place, there must be an assured supply at the input end of the facility and an assured demand at the output end. The projects tend to be highly sensitive to below-capacity operations, so that any implied constraints of either supply or demand are likely to make such a venture unfinanceable. In the second place, the finance for such projects, as indeed for any project, will comprise both equity and debt, and consequently the difference in value between output and input, after deducting all cash costs, must be sufficient to service and retire all debt financing within a reasonable period and must also provide a reasonable return to the equity holders. The meaning of the word reasonable in this context requires further definition.

The equity element represents the risk portion of the finance and will be entirely lost if the project should fail or perform insufficiently well. Because this risk exists, and because the risk may well be large, the return required by an equity-holder if the project is successful will also be large. However, provided debt finance is available at a cost below the project rate of return, the effect of borrowing is to raise the residual return to the equity investor to a level in excess of the project rate of return. If this residual return can be raised to the level of the equity investor's marginal requirement — as defined later in this section — then the project can be financed, but not otherwise. The financing objective for such projects is therefore frequently defined as minimising the equity requirement, and maximising the amount of debt which is to be raised.

Equity funds

In the heyday of capitalism, the major sources of equity funds were the savings of private individuals and, later, the profits generated by earlier successful equity investments. Today personal savings are largely held by pension funds and insurance companies, and entre-preneurial profits in many industries and countries are running at low levels in real terms and subject to heavy taxation. As a result, these traditional sources of equity investment are severely reduced in importance and, to the extent that they are now institutionalised, they tend to flow preferentially to a number of everyday, run-of-the-mill, relatively small scale investment situations, where the law of averages offers some protection against disaster, rather than to a single large one-off project where no such statistical protection is offered. The individual investor, applying his own savings to the best of his own ability, rarely has sufficient funds to achieve a statistical spread of risk and therefore is in a position to judge all projects on their merits. His own position does not prejudice him against large projects, as is the case with the institutional investor.

Because of these long-term changes in capital markets, equity funds for major projects of the kind we are considering are frequently in exceedingly short supply.

The equity investor must always reckon with the possibility that his investment will be lost. In order that he should be persuaded nevertheless to make the investment, he must also be satisfied that, if the project is successful, he will receive a relatively large reward. This reward will comprise three separate elements:

— the expected future rate of inflation in the currency concerned (in order to maintain constant purchasing power);
— a basic interest rate (to compensate for the investor's loss of use of the funds while they are invested in the project);
— a risk reward (to compensate for the possibility that the funds might in fact be lost altogether).

Of these three, the first and last are variable and perhaps subjective, but the second is in principle fixed for all currencies and all times. At a time, long ago, when there was effectively no sterling inflation, and when the British Government was considered an absolutely prime credit, it was able to issue open-ended debt at an interest rate of 2.5 per cent per annum. This rate may be considered to be the return required in real terms for loss of use of funds, there being no other factors at that time to be compensated for in the rate. It appears to be a reasonable assumption that a similar rate would meet a similar purpose today. To it must be added the first of the three factors, the expected inflation rate. The fact that open-ended British Government debt is, at the time of writing, yielding up to 15 per cent per annum, and that this rate includes the above basic 2.5 per cent interest component, and possibly an element for the fact that the credit is no longer universally considered to be risk-free, implies that the expected sterling inflation rate is around 12 per cent per annum. The third factor, the compensation for the risk of failure of the project, is of course entirely subjective and will vary according to the investor and the project. Overall, however, it is clear that the combined return required by the investor could well amount to 20 per cent per annum even in a stable currency with little inflation, whereas in sterling the requirement could well amount to 30 per cent.

Returns of this order of magnitude are difficult to obtain, and at the planning stage for a major project it is common to find that the project itself is unlikely to provide such returns. Debt finance must therefore be obtained at a cost below the project rate of return in order to raise the residual return to the equity investor to a level equal to or greater than his marginal requirement as defined above.

Debt finance

Debt finance is fundamentally different from equity in that a lender expects to be repaid irrespective of the success of the project for which his loan is used. The lender is in principle a bank, using funds deposited with it by its own customers. Its continued commercial existence depends upon its ability to repay such deposits and it therefore cannot and will not use the funds available to it to make loans in circumstances where the repayment of such loans on their due date is subject to excessive risks. Banking failures are the result of neglect of this basic rule, and although there is clearly room for some divergence of opinion as to which risks are excessive, unsecured project loans for major pipelines and refineries must usually fall into this category.

To attract debt finance, therefore, the lender must be satisfied as to the source from which he will be repaid in the event that the project fails to perform as planned. In the first place, he will wish to ensure that the cash flow from the project itself will be applied to repaying his loan rather than being passed to the equity investors. In the second place, he will wish to satisfy himself that other cash flows from other sources exist and are available to him to cover any shortfall in the project cash flows. Normally the only such cash flows arise from the other assets of the equity investors, and are made available to the lender in the form of direct guarantees from the equity holders. It is, however, feasible in certain circumstances to dedicate specific cash flows rather than provide general guarantees.

As a final fall-back position, the lender may well also require protection in the event of failure of all the cash flows of the equity holders; he may wish to be able to sell project assets in order to realise funds to repay his loan. Such a step would obviously be taken only in the last resort, and the borrower will require to be protected against any attempt to do any such thing at any earlier moment. In any circumstances which would permit the step to be taken, therefore, it is likely that the assets themselves will command a price in the market well below their optimum level, and quite possibly well below cost.

Since the lender will have protection in these various ways which are not available to, and are often directly at the expense of, the equity holder, he will expect a reward which is substantially less. In principle, it will comprise the expected future rate of inflation in the currency in question, together with an element in respect of the loss of use of the funds for the period in question, as well as a further element in respect of whatever the risk may be that the funds might be lost. In this theoretical sense, therefore, the lender's requirements are identical to those of the equity holders. In practice, however, the first two will be combined in the rate which the bank must pay to its own depositor, together with a possible increment forming part of the third element in the event that the bank itself is not considered in the market to be an absolutely undoubted credit. The bank will then calculate the rate at which it is prepared to lend into the given situation at a margin over its own cost of funds; this margin being related to the perceived quality of the security and the guarantor, rather than to the project itself. It is thus clear that in all practical circumstances debt finance will be substantially cheaper than equity.

Types of debt financing

We have now reached the conclusion that the fundamental objective in most pipeline or processing plant projects is likely to be to maximise debt and minimise equity. Within this objective it is then necessary to approach the various potential lenders in strict order: the lowest cost lenders first, to obtain the maximum from that source; then the next lowest, to obtain the maximum there, and so forth. The size of the debt requirement for major projects is such that it can often be met only after drawing upon a very large part of the available market, and taking in some funds at relatively expensive rates which may well exceed the overall project rate of return. The interests of the equity holders thus depend very heavily on the skill with which the more attractive sources are utilised, and a bank experienced in this type of business is normally employed for this specific purpose.

Subsidised finance

The first port of call will undoubtedly be an export credit agency. The second will be another such agency, and quite likely the third and fourth will be to others. The nature of these export credit agencies, their roles and the terms on which they will lend, are set out at length in chapter 11. Here it is only necessary to point out that in favourable circumstances as much as 80 per cent of the cost of a project can be financed from such sources, at interest rates as little as half those available in the commercial market. To achieve such a result, however, it is almost inevitable that impeccable guarantors will be required, and the provision of guarantees will be regarded by most people as being similar in nature to equity investment. They will therefore require returns similar to those sought by the equity investor, but of course without the 'true interest element' in respect of the temporary loss of use of funds. Such returns may be beyond the capacity of the project, or indeed guarantors of the necessary standing may simply not be available. Inability to provide suitable guarantors is a frequent cause of breakdown in export credit negotiations.

More fundamentally, many projects are located in countries which are too rich, or too poor, or of the wrong political colour, and are for these reasons not acceptable to some of the national credit agencies. It may also be the case that for economic, technical or political reasons some of the components required for the project must be supplied either by the home country, or another country which is not offering any export credit assistance.

For all these reasons, therefore, it would be most unusual to find a project achieving its theoretical maximum of some 80 per cent export credit finance, and the balance which remains to be financed in other ways is often the greater part of the total project cost.

Export credit finance is provided at subsidised rates by the governments of countries wishing to export into the country where the project is located. These host countries themselves may also offer government-subsidised loans to assist the construction of plant and pipelines adjudged to be in the national interest. The form of such assistance varies widely from country to country, and in many places is the subject of *ad hoc* negotiation quite outside any general legislative framework. This is particularly true of the less developed countries. It is therefore not appropriate to try to define the possible arrangements any more closely here, except to point out that the host government is likely to be the next cheapest potential source of funds after the export credit agencies of those countries able to supply the required goods and equipment. It should also be borne in mind that where the arrangements are on an *ad hoc* basis, they may turn out to be more useful than export credits. In particular, the onerous guarantee requirements of the export credit agencies may not arise in the case of host government finance.

Commercial loans

When all the sources of subsidised finance have been exhausted, or for whatever reason rejected, then recourse must be had to the commercial markets.

Most countries in whose territories a major project is located would prefer that the necessary finance should be raised externally rather than in their own domestic markets, even if adequate domestic markets exist, which is frequently not the case. Such external borrowings, if applied to the purchase abroad of goods which are then imported into the host country and incorporated in the project assets, have no effect on the balance of payments of the host country. If they are converted into the currency of the host country and applied to the purchase of domestic goods and services, the balance of payments effect is favourable. Other countries not involved with the project, but which possess domestic financial markets capable of meeting the needs of major projects of the kind we are considering, view the same balance of payments calculations from the other side, and with a jaundiced eye, so that it is frequently impractical to use such markets for the purpose.

As a result, the first approach to commercial markets is normally to the eurocurrency markets, which are in principle external to all countries. These markets are examined in detail in chapter 13, and therefore no such examination of them need be made here. One must, however, remember that they are almost always important, and frequently vital, to the financing of major projects all over the world, and that they may well be able to provide funds in the form of bank loans, bond issues, convertible stocks or even straight equity. It is from this source that the final tranche of major project finance is usually provided.

The eurocurrency markets are similar to the export credit agencies in that they too will normally require guarantees. This characteristic is shared by domestic capital markets, to the extent that they are available sources at all, but this will usually only be the case if either the project, or the guarantor, or the borrower (and ideally all three) is located in the country of the market. In such circumstances very large sums can be raised in debt form in major domestic markets, not only from banks, but also from insurance companies, pension funds and other investment institutions. Such institutions, with an assured inflow of funds and fixed long term liabilities, may be prepared to lend for longer periods than banks and at a fixed interest rate. Their involvement, where possible, can therefore be extremely useful.

Project finance

We have now reached the stage in the financing of the project where the basic steps have been completed. The decision was made to maximise debt and minimise equity, and to maximise the utilisation of low-cost debt. Export credit agencies were approached and debt raised from them against guarantees provided by the equity holders. The host government was approached, and low-cost finance arranged there (perhaps with guarantees). Eurocurrency markets were approached, and classical finance arranged there (against guarantees) and some finance (also against guarantees) was obtained from domestic capital markets. In many cases, the finance for the project will now be complete, but sometimes this cannot be achieved, usually because the original equity holders are either unable or unwilling to provide the required guarantees.

Alternative solutions to additional equity
The difficulty is the fundamental one of shortage of equity; the problem is how to attract investors to accept equity risks in the expectation of equity rewards. The difficulty is considerable, even though the contribution does not need to be in cash, but may be in the form of further guarantees.

The simplest approach to the problem is to bring in additional equity holders. If the pipeline or the processing plant were conceived by a single oil company, it may seek to bring in, for example, the host government as an equity partner, or another oil company, perhaps an end-user of the output, a supplier of the feedstock, or the contractor who will build the facility. Similarly, any existing consortium may be expanded.

Such approaches may attract some additional funds, but the scope is obviously limited. The number of potential additional partners is small and there are likely to be significant potential conflicts of interest among them. In some cases it may be possible to raise additional equity funds through existing capital markets — the mechanism for doing so is explained in chapter 12 — but this is only likely to be feasible if there is a substantial identity of interest between the new investors and the existing equity holders and if the project itself has a direct relevance to the new investors. These characteristics will not normally be present in the case of a project in a distant, developing country. On reaching this final stage, it may be worth yet another call on the host government, to try to wring further finance from them.

At this point the traditional financing mechanisms have been exhausted. The progression through the various possibilities is of course not a straight path, but involves loops, back-trackings and parallel advances. After all the possibilities have been tried from all possible angles, the finance is either assured, in which case the discussion is over, or it is not, in which case the project must be abandoned. Such an analysis would have been valid for most of the history of industrial investment. Because of the present shortage of equity, however, the analysis has led ever more frequently in recent years to an implied requirement to abandon major projects. As a result, a great deal of ingenuity has been devoted to finding ways of attracting essentially banking funds into basically equity situations, in order to permit the projects to proceed.

The resultant methods are generically covered by the term 'project finance', or 'project lending'. Both terms are also used in an extended sense to cover any finance for major projects, even if it is secured against sound guarantees which effectively classify it as entirely normal banking business. In the remainder of this chapter, however, the terms are used in the restricted sense of loans whose repayment is dependent upon the success of the project itself, whether it be a pipeline, a processing plant, or a wholly different venture.

The objective of all such methods is to reduce the risk to the lender of the project failing, by transferring to other parties as much as possible of such risks as do exist, and by then offering to potential lenders such returns as will attract them to bear such risks as remain.

The most serious project risks are failure to bring the facilities into working condition at all, delay in their completion, shortage of the input material at the right price, lack of demand for the output at the right price, and breakdown of the facilities or other interruption after start-up. The initial requirement is therefore to limit the extent to which these risks bear upon the lenders, without at the same time imposing specific financial obligations upon the borrower or his guarantors.

Risks associated with project completion

Taking these risks in order, failure to complete the project and delay in completion are different only in degree. The first aim here will be to have the contractor commit himself to supply a complete and working asset by a given date. In practice such a commitment will be difficult to achieve, but some approximation to it will usually be possible, given the probability that part of the risk can be passed down to sub-contractors, and that both main contractor and sub-contractors will be able to lay off part of the residual risk in the insurance markets.

Some of the completion risk, however, is likely to remain with the equity holder and the lender, and it may well prove difficult to agree upon the basis of its division between them. In practice, common ground may well be found along the lines that the equity holder commits himself to complete the facility, with counter-commitments from insurance companies and main contractors, who in turn have counter-commitments from insurance companies and sub-contractors. The pyramid may extend further downward still, but all these commitments and counter-commitments are likely to ignore the delay factor, leaving the lender to suffer the effect of postponement of his expected repayments. Thus the difference of degree between failure to complete and delay in completion may turn out to be the dividing line between the equity-holder's risk and the lender's risk.

An area for negotiation may be the question of who provides the additional funds normally required when delay is experienced. Sometimes the equity holder does so, but sometimes the lender finds himself in the unattractive position of being required to advance further funds in order to enable the project to be completed. He will normally seek to avoid such straits.

Assured supply and offtake

Early in this chapter the crucial importance of assured supply and assured offtake, both at the right price, were briefly mentioned. At this point we must examine the nature of their importance more closely, since once the plant is in normal operation they represent the most likely causes of difficulty for the project. In both cases, the preferred course of action is to have a company or government of the necessary financial strength commit itself to supply or receive stated quantities of material at stated prices.

Such commitments, like those of contractors and suppliers considered earlier, do not amount to full guarantees and do not therefore under present accounting rules have to be disclosed in the accounts of a company giving such commitments, and may for that reason be given rather more readily. In the case of a default under such a commitment, the remedy open to the aggrieved party will be to take action at law to recoup whatever damage he has in fact suffered. For financing purposes, therefore, it is essential that supply and offtake commitments be obtained from parties who not only can reasonably be expected to meet such commitments, but who are also sufficiently strong to meet any judgment for damages which might be awarded against them.

In practice, it is unlikely that a supply commitment can be obtained other than from an equity holder. The likely future direction of price movements for both crude oil and refined products is clearly upward, and a company or government which expects to have either available is likely to wish to maintain its freedom to dispose of its material where it chooses in order to maximise the advantage it can draw from the expected strong market. This view, however, may be subordinated to other considerations if the supplier is concerned to maximise his return from his material by having it processed or transported and sold in an upgraded form. A supplier holding such a view is likely to seek to give effect to it by investing in the necessary pipeline or processing plant, and the resultant combination in the one person of supplier and equity holder could well create the identity of interest needed to give rise to the issue of a supply commitment.

Alternatively, it is possible to envisage circumstances in which a supplier, not having an equity interest in a project, commits himself to supplying crude oil, refined product or other feedstock to the project, be it pipeline or processing plant, and also commits himself to take at least part of whatever comes out at the other end. Such an arrangement falls most readily into place in the case of a pipeline, where an oil company might well wish to use it simply to transport its material, and be allowed to do so by the equity holders, without taking an equity interest itself, provided that it offers both supply and offtake commitments against which the project can in part be financed. Similar arrangements could also be made in respect of a processing plant, where a company possessing the required input material and wishing to obtain the output material may contract both to supply and to offtake. However, although such circumstances are, in principle, feasible, they are in practice difficult to arrange and are unlikely to cover more than a small part of the capacity of the facility, since the bulk of it is normally needed by equity holders. It is therefore clear that the most likely partner for a supply or offtake commitment is an existing equity holder.

An offtake commitment may be structured as a 'take-or-pay' agreement. A tightly drawn take-or-pay agreement will stipulate that the committed party will accept delivery of defined quantities of material at defined prices, and will make payment for such delivery as defined, not only if the delivery is made, but also if for any reason whatsoever the delivery is not made. Such a commitment is as good as a formal guarantee, from the lender's standpoint, since it means that payments to the project, out of which the lender can be repaid, will continue even if the project fails to perform. From the standpoint of the party providing the commitment, the arrangement may be preferable to granting a guarantee, as it may be less subject to public disclosure. There will, however, normally be a cost penalty involved, to compensate for the additional complexity of the arrangement.

Such a tightly drawn take-or-pay agreement is a very great rarity. A more common situation would be one where a commitment has been given to accept delivery and to pay defined prices for defined quantities of material on delivery, but not to make payment in the event that delivery cannot be effected. In this way, a lender is protected against the risk that the borrower might be unable to dispose of his output at satisfactory prices, but not against the risk that he might not achieve the expected level of output.

An even weaker form of agreement which is sometimes seen provides for a commitment to accept delivery and make payment, but relates the determination of price to market levels at the time of delivery. Such an agreement is of little value as security for debt finance.

It is worth stressing at this point that lenders for pipelines and processing plants are most unlikely to accept any price risks whatsoever. Whereas for development and production projects lenders may on occasion be persuaded that the lowest foreseeable price which might be obtained for the hydrocarbons to be produced would still be sufficient to service and retire the relevant debt, in the case of downstream investments we are not talking of an absolute price level, but of a price differential between two links of the long integrated chain which stretches from exploration to final consumption. It is perfectly possible for the differential actually to be negative: for refined products to be worth less in aggregate than the crude oil from which they were manufactured, and for products to be worth less at the output end of a pipeline than at the input end. Such abnormal situations are related to particular patterns of demand and alternative use, and are usually temporary, but less extreme conditions, where the expected increment in value does exist, but not in a magnitude sufficient to cover the operating and financing cost of the facilities, are relatively common and can last for quite extended periods. For these reasons, lenders are likely to require that price risks be borne by equity holders, guarantors, suppliers or offtakers.

Breakdown or interruption after start-up

The division of risk between borrower and lender in the case of breakdown or interruption after start-up is likely to depend upon the nature of the causal event. Accidents and acts of deliberate sabotage are likely to be insurable, and since the main risk lies on the insurance company, lenders may well be prepared to accept the excess uninsured portion, unless limited capacity in the insurance market for this particular risk has caused the uninsured portion to be unusually large.

Even where the loss is covered, however, the profits lost as a result of the breakdown may not be recoverable. If the loss of profits is sufficiently severe to jeopardise the ability of the project to meet its debt obligations, then an agreement has to be reached as to whether the equity holders will make good any such shortfall, or whether the lenders must stand the loss.

The risk that the project might be interrupted by political action of the host government is one which may or may not be insurable, according principally as to whether the insurance market itself is to a significant extent dependent on the same political entity. If it is, it is likely to be unwilling to be seen to mitigate the effects of political actions by its own government. Lenders, perhaps surprisingly, may take a broader view. International banks are, almost by definition, beyond the control of any single government, except perhaps that of their home country. Most governments, also, are reluctant to act directly against the international financial community: they may want to borrow money themselves one day. For these reasons, lenders may well accept the political risk, believing it to be fairly small, while borrowers may well feel that the risk of unfriendly political action is actually reduced — to their own advantage — by virtue of the fact that the lenders would be the prime sufferers from such actions.

There may also be other causes of plant breakdown which are not covered by insurance. The same general questions will arise, but in much more acute form. A possible resolution of the difficulty might be that the equity holder commits himself to re-complete the project after breakdown, on the same basis as his original completion commitment, and protects himself with whatever insurance policies and counter-commitments he can obtain, while the lender accepts that his debt will take longer to repay. Other bases of agreement are, of course, also possible.

Cost of project finance

Discussions directed at obtaining genuine project financing frequently break down on the lenders' unwillingness to accept risks which the equity holders wish to shed. If we assume, however, that agreement is reached on all the vexed points we have briefly examined, then there is still the all important matter of cost to be resolved.

The lender will expect additional remuneration in respect of those risks he has accepted which are normally reckoned as equity risks. If they are relatively few in number and small in importance, the remuneration may be agreed in the form of an interest rate rather higher than would otherwise be the case. If on the other hand they are numerous and significant, the remuneration is likely to comprise both an interest rate above the normal and also a participation in the profits of the project. It is of course reasonable that someone participating in the equity risk should also participate in the equity rewards, and similarly reasonable that, since only part of the equity risk is borne, the reward participation should also be partial only.

On this basis if, for example, 10 per cent of the finance is provided in the form of genuine project loans, the lenders would, in principle, expect to receive less than 10 per cent of the residual equity profits. While it is theoretically conceivable that the equity holders should concede a share in the equity profits in excess of that indicated by the proportion of the capital contributed by the project lenders, it is in practice rare, since the normal effect of such an arrangement would be to reduce the residual return to the equity holders below the minimum acceptable to them.

The actual quantification of the reward to the lenders is likely to be the subject of detailed negotiations. These will not only be based upon the expected pattern of events if things go well, but also on the possible results of things going wrong. In economic terms, there will be a base case, and several sensitivity studies which will examine the effect of changes in the key variables. Those may include the start-up date, the capital cost, the eventual capacity of the installations, the operating cost, the volume of input material available, the demand for the output material, and the prices of input and output materials. In the nature of things, most of the possible variations from plan will be adverse, and the reward to lenders will reflect the effect of the possible disasters which might befall.

Documentation

There is likely to be a great deal of legal documentation to support the financing arrangements required for major projects of the kind we are considering. The purpose of the documentation is to ensure that the parties have reached agreement on all necessary matters, and have set down that agreement in a form which permits no misunderstanding. Whatever might come to pass, the respective rights and duties of the parties should be already settled, so that no further negotiation is needed.

In practice, this ideal may not be achieved, since events may come to pass for which no provision has been made. However, one area, which receives great attention and is usually satisfactorily resolved, is the question of defining an act of default and the rights of the lender if a default takes place. The definition of the act of default tends to be specific to the circumstances of the transaction, apart from fundamental general matters such as liquidation, or failure to make due repayments, but the methods used to preserve the rights of lenders do have certain common characteristics. In principle, the lender's aim will be to obtain immediate access, with as little trouble as possible, to a source from which he can exact repayment. He will not wish, for example, to be forced to proceed at law for damages, nor will he wish to prove his case as an ordinary creditor in a liquidation.

The lender's first line of defence will therefore be to require the assignment to him of rights over an ongoing income flow, which he can exercise only on the occurrence of an act of default. In project financing terms, this is likely to mean the assignment of supply and offtake contracts, and perhaps of certain consents, permissions, leases and licences necessary to the operation of the facility. Similar assignment of benefits under insurance policies is almost certain to be required.

The second line of defence will be in the form of arrangements to ensure that no other lender can obtain a priority position. In principle, this is usually achieved by taking out a first mortgage on the project assets. Such a step may also have the secondary result that the lender can sell the assets and recoup his loan out of the proceeds, but he will probably take the view initially that any circumstances likely to lead to a default situation in which he would have this right are likely also to result in the assets being of negligible value, so that his requirement to have a mortgage is primarily to ensure that no other party can sell or operate the facilities, rather than to realise funds for himself.

To cover this final requirement of exacting repayment in the event of non-performance by the project, the most satisfactory route is to have guarantees from parties of undoubted credit. In project financing, of course, such guarantees are, by definition, not available. However, it may be possible to obtain limited guarantees in support of covenants and other commitments provided in connection with the financing. The lender's course of action in a default situation would then be under the guarantee, instead of claiming damages at law for breach of covenant.

It should be borne in mind that the extent and nature of the legal arrangements required in a major project financing operation are such as to represent in themselves a significant cost element.

Conclusion

It is clear that there are no magic formulae to be applied to the financing of pipelines and processing plants. The fundamentals will be the same here as for any other projects: shortage of equity funds, need to maximise loan funds, need to maximise use of subsidised loan funds of various kinds, and possible inability to raise the total required without moving outside normal loan arrangements and turning to project finance of various kinds. Such arrangements must be negotiated in detail on an individual basis, and the negotiations may well be extremely difficult.

8

Liquefied Natural Gas Systems

'Natural gas' is a blanket term for a number of mixtures of naturally-occurring hydrocarbon gases, with methane being generally the most important component. Natural gas may be found either in association with crude oil (as a gas cap or in solution) or alone (non-associated) and sometimes far distant from crude oil production.

In the early stages of the development of the oil industry, natural gas present in the oil fields was considered to be a nuisance as it had to be separated from the oil. It was generally flared (burned) at the production site. A night-time flight over the oil rich Middle East is still filled with a strange glow emanating from the numerous gas flares in these oil fields. Today with the growing concern in the world relating to anticipated energy shortages, substantial quantities of natural gas are being gathered and distributed as a prime source of energy particularly in industrialised countries. The major problem encountered with the marketing of gas in production areas distant from the marketplace relates to transportation.

Transportation of natural gas

Gas pipelines have been successfully used for over land transportation, and many of the developed countries of the world that have gas production are interlaced with such pipelines that transport the gas with a minimum of treatment and in its natural state.

Transportation of natural gas over long distances, difficult terrain or over water requires the processing of the gas to condition it into an adequate form for such long distance transportation. Two methods are currently employed: the first involves the chemical combination of the gas with water to produce methanol (alcohol); the other method is to convert the gas into liquefied natural gas (LNG).

The conversion of natural gas to methanol simplifies the transportation problem as methanol is a liquid at normal temperatures and consequently can be transported in more or less conventional oil tankers. However, it requires large capital investment in processing facilities at the point of production and the use of the end product is restricted primarily to a chemical feedstock or an industrial user. It is technologically feasible for utility users to reprocess methanol into synthetic natural gas, but the economic viability depends mainly upon the utility being sufficiently distant from the producing area to benefit from the lower transportation costs.

The liquefaction of natural gas is currently the mechanism most widely utilised for the transportation of natural gas by sea. The gas is converted into liquid form through a processing plant that reduces the temperature of the gas to approximately $-260°F$ ($-162°C$), thus reducing the volume by 625 times and also eliminating the danger inherent with the volatility of natural gas. Such gas in liquid form is maintained in specially insulated storage tanks either on land or in specialised ships — LNG tankers — under relatively low pressure with the temperature being maintained by allowing a controlled evaporation known as 'boil off'.

Development of LNG plants

The technology for the liquefaction of natural gas has been in existence for more than forty years and has been extensively modified and perfected during that period of time. Initially liquefied natural gas plants were relatively small and were primarily 'peak shaving' facilities. The concept of peak shaving was to extract natural gas from a pipeline system during periods of low demand and to liquefy that gas so that it could be stored and made available for local distribution during periods of peak demand by regasifying the product. In this instance the major consideration was the reduction in the storage volume by 625 times from that amount required if the storage was to be made with methane being in its natural gaseous state.

Large-scale LNG plants designed to operate continuously throughout the year are called 'baseload plants'. The first baseload plant on a large scale was erected at Arzew, Algeria and has a rated capacity of 150 million standard cubic feet per day of LNG production. It incorporates storage facilities and an on-site marine terminal. The construction of such a plant became technically feasible when developments for the containment of the LNG on a sea-going vessel were developed.

All vessels used for the Arzew project were — and still are — designed with self-supporting tanks, i.e. Conch and 'Jules Verne' systems, and were the commercial application of the Conch/British Gas Corporation trials with the experimental LNG tanker *Methane Pioneer*. The next generation of LNG tankers utilised a 'membrane' system of LNG storage on board. With this containment system, the structural design of the ship also became part of the cargo hold with a thin metal 'membrane' separating the LNG from the insulating material.

Since the inception of transportation of LNG by ship, significant progress has been achieved in containment and insulation technology which has led to numerous competing systems, with LNG tankers now being produced with a capacity of 125,000 cubic metres of LNG storage. Current trends seem to give preference to spherical or prismatic 'free standing' storage tanks installed in fairly standard double hull ship construction. The boil-off from the LNG stored on board is not lost but first used to maintain the $-260°F$ ($-162°C$) temperature and then captured and utilised as a source of energy for the ship's power plant.

Cost of typical LNG system

The amounts of capital required for an integrated LNG project will vary according to the production capacity of the basic LNG plant. Let us therefore assume an LNG plant with a capacity of 1,000,000,000 (1 billion) standard cubic feet per day (scfd) of LNG production. The gas field will be located in the Gulf area of the Middle East, the wells will be deep and difficult to produce, there will be no infrastructure available therefore necessitating the project to be fully self-supporting. In our discussion we will review the range of costs and accumulate them on a roughly mid-point basis. We will also assume a fairly clean gas from wells of high production capacity with a minimal amount of liquid hydrocarbon content.

The investment required will logically be divided into five broad categories, namely:
   — drilling costs and production facilities;
   — LNG liquefaction plant and on-site storage;
   — infrastructure facilities;
   — LNG tankers;
   — receiving, storage and regasification facilities.

Drilling costs and production facilities
In order to produce 1 billion scfd of LNG sufficient wells will be required to produce between thirty and fifty per cent more gas, in order to allow for the extraction of the inert gases from the gas and to allow for utilising a portion of the produced gas as the basic source of energy for the operation of the plant and the support facilities. We will therefore need to allow for the cost of drilling sufficient wells to produce an upper limit of 1.5 billion scfd of raw gas, and for the pre-treatment facilities associated with production which separate any sulphur or sulphur dioxide, remove inert gases and extract any liquid hydrocarbons that may be present in the produced gas. If the gas field were an extremely good producer, between fifteen and thirty wells would be required with costs that could reach or exceed

$15,000,000 per well. If we assume a requirement of twenty-three wells at $12,000,000 per well, we would have an investment requirement of $276,000,000 for the wells, and the production facilities would range in the $62-94 million category and for our purposes we will assume $75 million.

Thus the total budget for this area of expenditure is $351 million.

LNG manufacturing plant and on-site storage

The manufacturing plant to produce liquefied natural gas with a total production capacity of 1 billion scfd would probably be designed with five 'trains' of LNG production, each with a capacity of 200 million scfd of LNG. This configuration would be designed in order to establish a balance between the benefit of large scale production and the risk of lost production through shutdown or repairs. In addition, there would be the requirement for five on-site cryogenic LNG storage tanks capable of holding 100,000 cubic metres, a power plant, a loading and unloading site for the LNG tankers and administrative facilities. The cost of such a plant would vary in accordance with the location, configuration of the gas and other variables but would be in the $1,250-1,875 million range.

Thus the total budget amount for this category is $1,565,000,000.

Infrastructure facilities

It is a fact that the major gas fields available for future production of LNG are located in areas of the world that have not enjoyed years of development and industrialisation and therefore lack the infrastructure normally associated with industrialised countries. It is therefore necessary to provide capital for the creation of housing, schools, roads, recreational facilities, and other amenities, which, for a project of this size, would approximate an investment of $50-75 million.

Thus the total budget amount for this category is $63,000,000.

LNG tankers

The technological achievements to date for LNG tankers have established a present standard of 125,000 cubic metres storage capacity. Such tankers have a cost of $125-150 million each and the number of tankers required will vary in accordance with the distance between the LNG plant site and the ultimate delivery point of the LNG. We can estimate that for every thousand miles of transportation distance 1.6 tankers will be required for a project of this size. Assuming an average distance of 9,000 miles, fifteen tankers would be required, with a capital cost varying between $1,875-2,250 million.

Thus the total budget amount for this category is $2,063,000,000.

Receiving, storage and regasification facilities

At the destination point of the LNG it will be necessary to construct an unloading pier, LNG storage facilities and regasification facilities with a cost of $300-450 million. Assuming two distinct market locations — an important consideration for financing risk — each location will require an investment of $300-450 million.

Thus the total budget amount for this category is $750,000,000.

Summary of costs

Drilling and production	$351,000,000
LNG plant and on-site storage	$1,565,000,000
Infrastructure	$63,000,000
LNG tankers	$2,063,000,000
Receiving and regasification facilities	$750,000,000
Total	$4,792,000,000

These figures have been developed on a 1976 cost basis. (1976 cost basis assumes engineering and construction beginning in 1976 with escalation factors included for the 4-5 years of construction time.)

Evaluation and feasibility studies

The drilling and production facilities as well as the infrastructure may be shared with crude oil production, if the project is concerned with associated gas, but the rest of the investment is specific to natural gas. The magnitude of the investment and the single purpose nature of the project has necessitated treating each project as a near self-contained entity; in other words a closed system for the purpose of financing and execution.

For these reasons a carefully detailed evaluation must be made of the system as a totality to establish commercial and economic feasibility before any financial commitments can be made. Risks to be evaluated can be broken down into the following major categories:

— political risk
— managerial risk
— technological risk
— raw material risk
— transportation risk
— market risk
— legal risk
— substitute energy risk

Political risk

A project such as the one described above will possibly include involvement, on a direct basis, from two or three different countries, namely gas producing countries and LNG importing countries; as well as a number of other countries on an indirect basis, primarily countries exporting technology and equipment to the production site. A careful evaluation of the political relationships among all of the countries involved must be undertaken before the project can be approved for execution. One important consideration is the stability of the existing governments and their mutual inter-relationships.

Because LNG plants are highly specialised and the equipment may not be readily available from alternate sources, the political relationships between the countries sourcing technology and equipment, and the gas producing countries must be evaluated during the construction period. A key component to the LNG plant not delivered for political reasons could jeopardise a $5 billion investment. The feasibility study would have to consider those critical items that are not readily available from multiple sources so that political considerations can be properly evaluated.

During the production cycle, which would probably run for a period of 20-25 years, the political relationships between the LNG producer and the LNG buyers will be essential. Evaluation and forecasting is more difficult due to the length of time involved. There is, however, an important consideration in that LNG is not at present a commodity such as oil. The producer of LNG is restricted in delivery points by the number of receiving terminals in existence in the world. In most instances the total investment associated with a project of this magnitude would tend to stabilise political relationships due to the chaos that would be evident if a $5 billion project failed.

Managerial risk

The management structure for the LNG project will have to be carefully organised and selected. This will be one of the more critical aspects of the project as an LNG project of this magnitude immediately becomes a major operation with very substantial assets to control and protect. The problem is further complicated by the tendency toward remote locations for gas fields, which will require special incentives to attract the managerial calibre required for the orderly running of the enterprise. Because of this, LNG projects have been —and may increasingly be— reflected as 'joint ventures' with established business enterprises and state agencies from producing countries. Thus outside management skills and techniques can be introduced into the LNG project. While such joint venture partners may not risk their corporate assets directly on behalf of the venture, major corporations have established a reputation that must be protected and will ensure adequate management, attention and assistance for the orderly operation of the LNG project

Technological risk

Throughout the history of cryogenics there have been active developments of new processes and technology. Recent developments have established new levels of efficient operation and reliability. All such efforts have not, however, been successful and technological evolution continues. This is a natural phenomenon in any area of technological development, but, with substantial quantities of capital at risk, particular attention will have to be paid to the technologies to be utilised, both as relates to LNG process and mechanical equipment selection. Operating efficiency, in terms of energy consumed per unit of output, will not be as critical for such a project as will be the absolute reliability of the process, the equipment, transportation system, etc. The plant will have to be designed with sufficient excess capacity to meet deficiencies in production through shutdowns, maintenance and/or low levels of performance in order to keep the complete system in balance. Should everything function well there will always be ample opportunities for extra shipments to be made which will merely enhance the overall profitability of the project. Prior to funding being available, independent consultants will be required to make detailed reviews and evaluations not only of each aspect of the project but also of the project as a whole.

Raw material risk

The geological work required to establish the existence of sufficient producible gas in predetermined quantities will have to be carefully evaluated and double-checked in order to ensure that there are sufficient reserves in place for the plant's requirements. Because of the large amounts of capital involved, necessary safety margins of excess reserves will have to be established at higher levels than would be considered normal for a more moderate investment. Such geological consideration may require the drilling of excess or standby wells for safety margin reasons, as well as carefully designed production systems, in order to ensure that a problem with one particular well does not have a significant effect on the production from the other wells. Such margins of safety will vary because of many factors, but they may have to incorporate 'overcapacity' of up to 50 per cent for certain key aspects of the project.

Transportation risk

LNG tanker technology has improved substantially during the past decade, and additional improvements will probably be reflected by the introduction of more efficient containment systems. The power plants for the ships will, of necessity, be of a standard, and time-proven, configuration — as will the overall hull design as well. The detailed calculation of tonnage requirements will have to be established on a conservative basis as a breakdown of a single LNG tanker would have a significant impact on the cash flow of the project. This would affect all parties concerned. One must also consider 'down time' for periodic maintenance and the fact that, as the ships age, the cruising speed will decline, thus eventually increasing the requirement for LNG tankers.

Market risk
There are few gas utility or gas-utilising industrial corporations whose credit worthiness alone would support the capital requirements of projects of the kind we are discussing. The LNG will be sold for a specified number of years in predetermined annual and monthly quantities on a 'take-or-pay' basis. Financiers will have to look through such a purchaser of LNG to its ultimate customers. Lenders will have to be satisfied with the long term prospects of strong market requirements. In the case of a utility purchaser, the major risk would be economic dislocations within their market area which could substantially reduce the demand for energy. In the case of industrial purchasers the company's own technology may become obsolete, thereby reducing its output and energy requirements. In certain instances the government of the LNG purchaser may be required to guarantee the 'take-or-pay' contract in order to make it 'bankable'. This will depend on historic data relating to exchange controls, balance of payment criteria and government intervention in industry.

Legal risk
Long term contracts between parties governed by different legal systems are fraught with constantly recurring difficulties. Adequate attention to arbitration arrangements in the various agreements is therefore an essential ingredient in minimising legal risks.

Substitute energy risk
An LNG plant and system is relatively inflexible and basically designed for single purpose operations. There is a risk, admittedly small, that a technological breakthrough could occur, resulting in compatible energy sources and at a substantially lower cost than delivered LNG. Such a phenomenon may provide sufficient incentive to the purchasers of LNG to breach their take-or-pay contracts which would have a devasting effect on the LNG project. This would be particularly possible if such substitute form of energy could be provided within the LNG purchasing country or non gas-producing country that would lack the incentive of protecting its LNG industry and therefore encourage the breaking of LNG contracts. This risk could be mitigated through government guarantees and/or government sponsored insurance protection.

Financing of LNG systems

LNG projects are too capital intensive for standard 'balance sheet' financing and hence recourse will have to be made to project financing. This means that these projects will be financed primarily on a 'debt basis' as previously discussed in chapter 7. As such, shareholders' equity will be minimal, and the supporting criteria for the debt structure will rest on the underlying contracts, equipment, technology and political considerations and will be supported by appropriate contracts for raw gas, ocean transportation and LNG take-or-pay sales contracts.

The financial structure of an LNG system such as we are discussing may take several forms. The total system may be structured within a single corporate entity, but more than likely will be separated into three different companies, some of which may be joint ventures with business enterprises or state agencies:

— the LNG plant and related production facilities;
— the transportation system;
— the importer-owned receiving terminal.

It is evident that the financing arrangements for the three distinct phases of an LNG system must be closely coordinated as well as the construction of the various facilities required.

Due to the very heavy cost of borrowing such substantial quantities of funds timing is critical in order that all phases are available for full scale operation at approximately the same time. The financing institutions will also require that escalating costs, relating to operating expenses for the LNG plant and the LNG tankers, be passed on to the LNG purchaser so as to insulate the necessary cash flow for debt servicing. As the major component of the delivery price of LNG is the amortisation of the capital cost of the facilities — which will have been clearly defined prior to the beginning of production — escalation of operating costs is generally acceptable to the LNG purchasers.

An economic model will be prepared for financial analysis, a programe of 'sensitivity analysis' will also have to be created. The sensitivity analysis will have to include the impact of escalating capital costs, increase in interest rates, startup delays, less than 100 per cent output capacity and temporary shutdowns. The debt service requirements must be sufficiently protected in the basic project criteria to compensate for variables such as outlined above.

The LNG plant
The LNG plant, as we have determined, will require in excess of $1.5 billion of capital. The major portion of such capital requirements will be generated through the issuance of debt. Since most of the

technology, engineering services and equipment procurement will be imported from major industrialised countries which could, and often will include the country purchasing the LNG, the financial structure will maximise the utilisation of official export finance programmes. With a level of exports for the project exceeding $1 billion over a four to five year period, there will be substantial incentive for the major exporting countries to establish special export facilities for a project of this nature.

Approximately two-thirds of the capital cost can be financed through such export finance programmes which generally allow for longer repayment terms and lower interest rates than are generally available in the private sector financial markets. The remaining requirements will be in part supported through equity contributions and/or financial guarantees of the shareholders, and will be supplemented with direct borrowings through international banking syndicates.

In order to calculate debt servicing requirements based on different amortisation schedules for the various sources of funds, a financial model which will measure the capital cost will be utilised. Normally such a project would be expected to amortise fully its debt during the first half of its scheduled life; the cash generated during the second half would be available for shareholder dividends. A profit sharing scheme may be worked out between the sponsors of the project and the lending group. Such a programme could create sufficient incentives to generate the 'imaginative banking' required to achieve the level of funding for the project.

In order to establish a minimum sales price for the LNG that will maintain a positive cumulative cash flow for the LNG plant, debt servicing requirements will have to be combined with operating costs and raw material costs and converted to a cost per million Btu. Due to the high capital cost, it may be necessary to establish a declining price level for LNG and/or advance payments from the purchaser during the early years of production when financial costs are greatest. We can assume that official export financing will be available with twelve-year repayment terms beginning after commercial production has commenced; however, under current market conditions, the direct private sector loans would probably not be available for a period in excess of five years. (When combined with a 4-5 year construction cycle, this generates an average life of 5-6 years for the private sector loans.) The permissive lending of the early 1970s on an international scale is not evident at present in the marketplace. The financial cost of capital during the construction period — which under present land-based concepts requires five years — represents a substantial portion of the capital costs. This period could be substantially shortened through the utilisation of new techniques, which will be covered in a subsequent section of this chapter.

The transportation company

The coordination and management of a $2 billion tanker fleet will be a critical factor in organising finance for these ships. As discussed previously in chapter 6, there are numerous specialised finance programmes available from official entities of countries that build LNG tankers which vary in accordance with the day-to-day shipyard conditions of the prospective shipyard to be utilised. These programmes may include construction subsidies as well as operating subsidies and, in some instances, require that the ship being financed services a country providing the finance.

During the permissive lending period which existed in the early 1970s a number of LNG tankers were built on a speculative basis, and financing was arranged based primarily on the owner's balance sheet. Projections were prepared showing the world demand for LNG; these indicated a severe shortage in LNG tanker tonnage availability. Such tankers were not under contract for specific LNG projects and the absence of the strict project criteria outlined in this chapter has caused a number of ship owners to experience severe financial difficulties. Estimates in excess of $25 million a year for the maintenance requirement for an idle LNG tanker are not uncommon. Nevertheless, a prospective LNG project might reduce its overall operating cost on a delivered LNG unit basis by taking advantage of the currently existing situation where hundreds of millions of dollars have been invested in tankers that are idle.

It will be extremely difficult to obtain financing from commercial sources for new LNG tankers while there is substantial excess tonnage lying idle. However, countries who have a strong dependence on shipyard activity for their economic well-being may continue to offer special incentives for the construction and financing of new LNG tankers in order to satisfy domestic political economic considerations.

The importer-owned receiving terminal

To date the trend has been for the LNG purchaser to establish his own receiving and regasification facilities. These would more than likely be established from the purchaser's requirements given in the first LNG purchase contract, with design allowances for future expansion as additional quantities of LNG may be contracted from diversified sources. Such users would either avail themselves of long term financing markets in their own country or seek credits from international financial syndicates. The likelihood of export credit being utilised for the erection of receiving terminals is remote as LNG importing countries would generally be sufficiently industrialised to have domestic capability for the construction of such receiving terminals.

Growth prospects in LNG

When considering the forecasted increases in energy requirements in the industrialised countries together with continuing environmental pressures, it is apparent that natural gas will continue to provide a significant portion of the world's energy requirements. During the next decade it is anticipated that LNG shipments will increase five to tenfold over the mid-1970s levels.

By the mid-1970s there were eight major LNG projects in production and under long term LNG sales contracts. These involve LNG exports from Algeria to the United Kingdom, France and the United States; from Alaska to Japan; from Libya to Italy and Spain; and from Brunei to Japan. All together these projects represent 1.5 billion scfd of shipments.

There are three additional projects under construction which will commence operations before 1980 and they will supply an additional 2.3 billion scfd of LNG under long term contracts. There are ten to twelve projects under active negotiation that will yield another 5 billion scfd in the early 1980s and an additional 5-10 billion scfd could be supplied from a further list of ten to fifteen identified potential projects not yet sufficiently developed to estimate start-up dates, but estimated to be on stream by 1985.

It is apparent that the amounts of capital required for future LNG projects will exceed $30-40 billion. Improved technology will have to be developed in order to reduce such astronomical capital requirements.

Presently the technological improvements are principally directed towards the minimisation of construction costs and time in erecting LNG plants and on-site storage facilities. In the example outlined in this chapter we reviewed the costs and timing requirements for LNG plants and storage facilities to be built on land at a site near the gas field with marine facilities. In the future, LNG plants may be designed to be constructed in a shipyard as a joint unit with storage facilities similar to those used for LNG tankers. It can therefore be constructed under controlled conditions, towed to the ultimate site and ballasted to rest on a preconstructed caisson providing permanent mooring. Such an approach could result in a net reduction of up to 20% of the investment required, including interest during construction. Furthermore, a shipyard-constructed LNG plant and storage facility would qualify for the official programme for export ship financing that would cover a greater percentage of the capital requirements than similar programmes for land-based facilities, thus substantially reducing debt service chargers and therefore the delivered price of LNG. By assembling the LNG plant and storage facilities in a shipyard, as much as twenty-four months could be eliminated from the construction time; original engineering work could be used for multiple applications and the inefficiency of field construction and erection would be minimised, resulting in lower direct capital cost, lower interest cost during construction and faster start-up time for commercial deliveries. The need for costly piers with cryogenic capabilities could also be eliminated.

The application of LNG processes in marine installations poses a number of questions regarding seaworthiness as well as production capability and reliability under the rocking motion that would be experienced in such a floating LNG plant. J F Pritchard and Company, a subsidiary of International Systems & Controls Corporation, and the pioneer in designing baseload LNG plants, has identified a number of potential applications where floating LNG plants would be feasible. Pritchard has been working jointly with Sistemas Navales Y Criogenicos SA (CRINAVIS) of Spain which operates shipbuilding facilities specifically designed for constructing cryogenic tankers as well as process plants and storage systems to be barge mounted for marine operations. The LNG seabed supported system L3S™ developed by the Pritchard-Crinavis joint venture is accommodated on a single barge and incorporates the Pritchard proprietary 'PRICO™' LNG process which is jointly owned with Kobe Steel of Japan and utilises Crinavis's own storage design. Moss Rosenberg, a Scandinavian shipyard, has also done work on marine-based LNG process plants and is actively pursuing such technological changes.

Beyond the next decade it is difficult to forecast the demand for new LNG facilities as such demand will depend upon world demand for energy and technological developments for alternative environmentally sound energy systems. We do not anticipate that in the three major LNG importing areas of the world (USA 40 per cent, Japan 33 per cent and Europe 27 per cent of total imports in 1975) natural gas will be supplied in the form of LNG in significant proportion to total energy consumption. Nevertheless, LNG will continue to be a significant 'supplementary' form of energy during the next two decades. For the LNG industry, even a modest increase in its share of the energy market will signify quantum growth.

PART D
TAXATION AND
INSURANCE ASPECTS

It is important that the raising of finance and its use within the business is done in a cost-effective manner. Two key areas which are highly significant in achieving overall cost-effectiveness of any business or project are the proper consideration of taxation and insurance.

The impact of taxation on financing decisions cannot be overstated. In a world of independent taxing authorities, the enterprise conducting business within many nations and across national boundaries faces an almost infinite variety of types of taxes, levels of tax burdens, tax incentives, patterns of tax administration and possible overlaps in tax systems. Taxes are an important variable in capital budgeting decisions, in choosing the form of business entity for foreign operations, in selecting the financing strategy for the group and in the formulation of remittance policies. Because of the pervasive importance of the tax variable in financial management and because of its diversity and complexity, companies involved in the international petroleum industry must examine the tax implications of all financial decisions if they are to minimise tax obligations within legal limits. The purpose of chapter 9 is to alert the non-tax specialist engaged in financial management to some of the more important taxation areas which should be taken into account.

Insuring against specific risks is invariably a prerequisite to concluding the raising of certain types of finance. This is especially the case in non-recourse project financing. Insurance provides a means of obtaining financial security against some risks which face every business enterprise. Chapter 10 describes the insurance business so that the reader will have a basic knowledge of this industry and its working practices. It then proceeds to highlight the kinds of risks which face the international petroleum industry and which are insurable.

In dealing with these highly technical and complex subjects in two short chapters it is only possible to provide an overview but which hopefully is sufficient to alert the non-specialist to the kinds of problems which might be encountered. At the end of each chapter references provide further reading should the reader feel the need for information in greater depth.

9

Taxation and the
Petroleum Industry

The international petroleum industry, of its innate nature, conducts business within many countries and across many international boundaries. The oil companies, therefore, face an infinite variety of fiscal problems embracing every kind of tax, every possible tax base, every level of tax burden, every sort of tax incentive and every conceivable variation in principle and in detail of application. Moreover, because of their national rather than international origins, tax regimes frequently overlap with one another across international borders, thus creating further significant complications — in short a scene of bewildering complexity.

The tax factor pervades every aspect of operations including, for example, the design of corporate structures particularly for overseas operations, the choice between share and loan capital, the policy regarding the declaration and remittance of overseas dividends, the flow of interest payments and the most profitable use of working capital resources to name but a few. Thus, because of the importance of the tax factor international operations require the exercise of considerable expertise if tax obligations are to be anticipated and minimised within the limits allowed by the law. As the complexities and sophistries of international taxation are obviously outside the scope of this book, the intention is to sketch in the scene as simply as possible and to try to alert the non-tax specialist engaged in financial management to some of the more important areas in which tax considerations should be taken into account, since to recognise and identify a problem goes a long way towards its solution. Thus this chapter is, in a sense, a commentary on the others.

In making management decisions, all relevant factors have to be taken into account — operational, financial and even political in some instances — and properly and expertly planned taxation must be accorded its due weight. Taxation, however, must always be the servant, never the master, of commercial requirements, since, if tax objectives are permitted to override commercial considerations, this usually leads to artificiality. In this event the hoped-for advantages may prove to be illusory. Indeed, in extreme cases, such arrangements can lead to political difficulties and the introduction of severe anti-evasion legislation, which can have unforeseen and damaging effects. This does not mean that tax planners should not use their ingenuity to attempt to reconcile apparent or real conflicts between operational and fiscal requirements, but the starting point must always be a genuinely viable commercial operation, which the tax planner endeavours to organise in the way that will be fiscally most advantageous when all considerations are taken into account. It is usually unfruitful to seek out a flaw in a tax regime and then try to develop a commercial operation to exploit it. Remember particularly that the petroleum industry is frequently the centre of political controversy, and in a climate of increasing nationalism, operations both physical and fiscal must be concluded in such a way that the company concerned is, and can be seen to be, a good citizen of the country in which it is located, meeting its fiscal obligations honourably. This will help to increase stability if unwarranted political pressures are brought to bear on it.

The approach herein will be to look at the tax situation associated with the activities of the international oil industry under the following headings:

- —the flow of oil — that is its production, transportation by pipeline or ship, its refining and its marketing;
- —corporate structure, equity capital and dividends;
- —loan capital, its servicing and repayment;
- —the flow of services between head office and subsidiaries;
- —miscellaneous matters.

Naturally these headings are not mutually exclusive, but the problems, will be considered under the headings into which they fit most naturally.

It might appear that by dealing with taxes on profits before those on dividends and interest the central theme of the book is being given a secondary place. It is felt however that the problems of corporate structure and the flow of dividends and interest can only be properly comprehended against an understanding of the taxation of profits. Indeed the form and structure of an overseas operation will significantly affect the manner in which the domestic tax burden of the shareholder will fall — that is whether the tax will be levied directly on the profits of the operation or on the dividends declared out of those profits. The granting of double taxation relief will be similarly affected.

The flow of oil

This will be divided into four main headings: firstly, royalties and other related payments; secondly, taxes levied on production profits only; thirdly, taxes imposed on profits from commercial and trading activities, generally including oil production, transportation by pipeline and/or ship, refining and marketing; fourthly, international complications arising out of the foregoing, including double taxation agreements and other international trading matters.

Royalties and other related payments
The simplest form of tax on production is probably the royalty — sometimes known in the oil industry as a severance payment or a barrelage tax. Such imposts on mineral extraction activities have existed since time immemorial, and they probably constitute one of the oldest forms of taxation. But that does not preclude asking the question — how much must be paid and when? The answer to such questions is usually comparatively simple when compared with the problems imposed by the more sophisticated imposts, but nevertheless difficulties do exist. These will usually include the actual measurement of the amount of oil or gas extracted, and its valuation within the terms of some rather esoteric formulae designed to take account of the differences in circumstances between one producer and another. For instance, royalty is often calculated as a percentage of the value of the oil or gas at the well head. Since, however, in cases where production is offshore, commercial values are usually established onshore, and royalty is calculated by reference to a market value which is arrived at by deducting the cost of transporting the oil or gas to land. This in turn requires a set of rules, sometimes very complex, in accordance with which such costs must be determined. If, for example, the transportation is by pipeline, and is used not only by the licensee who owns it but also by a third party, there will usually be sophisticated rules for allocating standing costs and variable costs, and sometimes for the sharing of interest on the capital employed. Clearly these are practical commercial/accounting matters on which it should not be too difficult to reach reasonable agreement.

Royalties are often, but not always, a charge against profits for the purpose of calculating taxes on income, including both those imposed on production profits only, and those imposed on profits from commercial activities generally. They may in some instances be regarded as a payment on account of the tax on production profits, as was the position in the Middle East for many years, but is much less common now. One of the most important points to bear in mind when considering royalties is whether or not they qualify for double taxation relief. Generally speaking they do not, so that any calculation of relief due — along the lines of that described in the later discussion of double taxation — must usually ignore the royalty, to the detriment of the party which has borne it.

There are sometimes alternative provisions under which the licensing authority may take royalty in kind rather than in cash. In such cases the precise wording of the relevant provisions must be studied, and its relationship and effect upon the basic royalty provisions must be established.

Other payments which normally have to be made by concessionnaires in return for the right to prospect for and, if successful, to produce oil and gas, are rentals in respect of the acreage worked and, in some cases, lump sum down payments, which can be either fixed or part of a competitive bidding system, under which the concession rights are granted to the highest or most favoured bidder. Typical examples of the latter include the very large payments made by certain companies to acquire acreage on the Alaska north slopes a few years ago, and the sums paid by various companies in the fourth licensing round in connection with the UK continental shelf. The tax treatment of these payments and their interaction varies enormously. In some instances rental payments are regarded as being on account of royalties which are reduced accordingly. Lump sum payments may be treated as an expense, or they may be written off against profits under some kind of amortisation formula. In some instances certain of the payments may not qualify for relief at all. The possibilities are endless, and any relevant documentation and legislation must be studied carefully in every case.

Taxes levied on production profits only
These are becoming more common. Until recently the best known were the so called 'fifty fifty' taxes in the Middle East under which the profits of oil production were divided equally between the government of the country where the oil was situated and the oil company operating the concession — although latterly the split has favoured the governments. In point of fact these taxes were theoretically taxes of general application, but they had such a high tax threshold — that is the profit level which had to be attained before any tax became payable — that in practice only the oil companies actually became liable to pay them. This was a device to ensure that the 'fifty fifty' tax would qualify for double taxation, since, in some countries where oil companies had their tax residence, and notably in the UK, a tax on income had to be a tax of general application if it were to qualify for double taxation relief; certain uncertainties of this kind hang over the UK Petroleum Revenue Tax (PRT). With the increasing degree of participation by the Middle Eastern countries in the exploitation of

their own reserves these 'fifty fifty' taxes are of declining importance.

Production taxes in existence outside the Middle East include those in Barbados, Malaysia, Nigeria, Norway, Thailand, Trinidad and Tobago, and the UK. Since the UK Petroleum Revenue Tax has attracted so much attention and is potentially of great significance, it is thought that it would be helpful if its main features were summarised.

PRT is a completely new concept in UK taxation. Although it is administered by the Inland Revenue, and is linked with, and makes use of some definitions contained in the main tax code, it is nevertheless a completely self contained and independent tax and does not form part of the main corpus of the UK tax code (embracing income tax, corporation tax and capital gains tax). Thus, to take but one example, a loss incurred by a taxpayer for PRT purposes cannot be used to reduce the corporation tax or income tax payable on profits made on other operations carried out by that taxpayer.

PRT is imposed only on the profits of production activities carried out in the UK and on the UK continental shelf, and does not apply to profits earned by agents who carry out the multifarious service and support activities for the licensees. The essential features of the tax are that it is imposed on each participant for each field on his assessable profits, as reduced by his allowable losses and his share of any oil allowance, but subject to an annual limitation. The concept of assessable profit, which is a very unusual one, is discussed below.

A participant is a licensee in a block where oil or gas has been found and is being exploited. Most exploration and production is carried out by consortia. These consist of companies linked together contractually for a common purpose in such a way that each retains its own individual, albeit undivided, interest in the operation, and is taxed independently on its share thereof. Each is a participant for PRT purposes.

The expression 'oil field' is used effectively in its everyday meaning of a location where oil or gas has been found and each is given a name; for example Beryl, Viking, Auk, Brent, Forties etc. The field by field concept is applied rigorously, so that, for example, there is no possibility, except in the special circumstances mentioned below, of offsetting profits in one field against losses in another, although losses in a field can be fairly freely carried forward and/or backward against profits from that same field. The field by field concept is broken only in the event of expenditure proving to be abortive because no oil or gas is found, and of the abandonment of a field which has ceased production. In such cases the participant can claim a deduction against the profits of his other fields.

Perhaps the most striking feature of the PRT is the concept of profits and losses. These are not calculated by taking the commercial profits and making prescribed adjustments thereto. For PRT one is required on the one hand to value the oil or gas won and saved, and on the other to itemise the admissible expenditure. These are two separate operations, each having its own rules and procedures. The difference is profit or loss. The valuation of production is made each half year, but expenditure becomes deductible only after it has been vetted and certified, which may be a considerable time after the incurring of the expenditure. For these reasons, and because of peculiarities in the rules governing the admissibility of expenditure outlined below, there will usually be no readily discernible relationship between the commercial profits and losses and those calculated for PRT.

One of the most fundamental principles underlying PRT is that it is calculated by reference to the open market value of oil and gas. The provisions which ensure that this is achieved are precise, extensive, and detailed, and they pervade the whole of the legislation. Any oil sold under contracts which do not satisfy the very stringent arms length conditions, and any oil delivered to a refinery for processing, must be brought in at open market value as defined in the Act. Oil sold genuinely at arm's length is brought into the calculation at its sale price.

Most of the expenditure incurred in finding and producing oil from a given field is allowable in arriving at the profit or loss, but no distinction is made between capital expenditure (that is expenditure incurred once and for all with the object of bringing into existence an asset having an enduring benefit or advantage for the trade) and revenue expenditure. However expenditure on assets which have a long, useful life and which may be used in more than one field, is shared out amongst participants on a suitable basis.

But there are two other even more important and unusual provisions concerning the admissibility of expenditure. The first is that expenditure on bringing about the commencing of the winning of oil, or the ascertaining of the size and characteristics of a field and on a few other related matters qualifies for a supplement of 75 per cent. That is to say the participant is entitled to claim as a deduction 175 per cent of such expenditure. The second, which is in a sense the *quid pro quo* of the first, is that no deduction whatever can be claimed for expenditure in respect of interest or any other pecuniary obligation incurred in obtaining a loan or any other form of credit — there is no exception whatever to this rule. Certain other expenditure, including that on purchasing an interest in a concession from a third party, does not qualify for deduction either.

One significant consequence of the unusual structure of the tax is that no tax is payable for a field until the value of production exceeds admissible expenditure, uplifted by 75 per cent where appropriate.

This feature was deliberately written into PRT so as to defer the payment of tax until the very considerable initial costs of finding and extracting the oil are covered from income. It is interesting that, in practice, this is not dissimilar in effect to the UK corporation tax system, although the result is achieved by different means. For corporation tax, once the company has commenced trading, most of the vast expenditure incurred up to that date on the exploration and construction of production facilities becomes tax deductible under one or other of the relevant tax rules and no tax is therefore payable until that expenditure has been recovered.

Three other matters must be mentioned if this brief summary is to be comprehensive. Firstly, the oil allowance, which ensures that the first 500,000 tons of oil from each field for each chargeable period of 6 months up to a cumulative total of 10,000,000 tons is exempted from PRT. The exemption is per field and is divided *pro rata* among the participants in that field. Secondly, the annual limit of tax to be paid, which by a complex formula prevents the PRT biting so hard that a participant's profits are reduced below an acceptable level of return on capital employed. This is also calculated for a given participant on a field by field basis. These two provisions together are intended to ensure that the taxation burden on small marginally profitable fields is not too onerous — otherwise such fields might not be developed. Thirdly, natural gas sold to the British Gas Corporation under contracts entered into before the end of June 1975, is not subject to PRT since it is felt that the prices for gas negotiated up to that date were unrealistically low.

The rate of PRT is currently set at 45 per cent and is expected to remain at that level in the foreseeable future, but if any change has to be made, it is anticipated that it will be dealt with in the annual Finance Act. PRT is deductible in computing for corporation tax purposes the profits derived from production operations. Its treatment for double taxation relief purposes is considered briefly below.

Taxes imposed on the profits of commercial activities generally
Taxes on profits which are of general application have a variety of names including income tax, corporation tax, industrial tax and federal tax, to name but a few — the term corporation tax is used herein to cover all these imposts in a general way. Although corporation taxes are imposed generally on profits from all commercial activities, which will normally embrace all aspects of an oil company's operations — including production, refining, transportation and marketing — there are many variations.

In particular, in territories which have a production tax, corporation tax may not be levied on production profits, but if it is, it is important to know whether the production tax is deductible in

146

calculating profits for corporation tax and vice versa or whether each is independent of the other. It is equally important to establish whether the corporation tax covers all oil activities in the same way. For instance, there may be special rules for dealing with shipping profits. Clearly, since the possibilities are endless, only a few points of principle can be mentioned here.

In considering any tax on profits, it is of fundamental importance to know what the scope of the tax is. Generally speaking, corporation taxes are levied on all the profits, wherever they arise — that is on world income — which accrue to companies resident in the territory in question. There are various definitions of resident in this context, but it commonly applies to any company of whatever nationality or wherever incorporated, whose policy is directed at board meetings held in that territory, as is the case under UK law — in such instances the 'mind and management' of the company is said to be in that territory. The USA is an exception in that companies incorporated there are effectively resident there, as they are liable to US tax on world income by virtue of their US incorporation. There are exceptions to the general rule about the taxation of world income of a resident company — for example, the profits earned by the overseas branch of an Australian resident company are not subject to Australian tax, as long as they are taxed in the overseas territory where they arise. For the purposes of the foregoing the residence of the owners of the shares, and the place of incorporation of the company, are generally not significant. For example, a company incorporated in the UK, having German shareholders and resident in the Netherlands would be subject to corporation tax only in the Netherlands. Thus it is important to distinguish clearly between these various concepts — that of residence being the important one.

Virtually all countries impose domestic corporation tax on the profits derived by overseas companies from branch operations carried out in their territory. A branch operation is one which is carried out in an overseas territory without the company concerned becoming resident in that territory or setting up a local company. The tax rate applicable is often different from that applied to resident companies. It can be higher or lower, but is frequently higher, to compensate for the fact that such branches do not usually declare dividends in the overseas country, and there is therefore no possibility of a withholding tax being imposed by the country where the operations are being conducted. Canada is unusual in that it imposes a tax on branch operations, which is to some extent a substitute for a withholding tax. The question of what constitutes the territory of a country for the purpose of branch operations is an interesting one, particularly in the context of offshore oil operations. Some countries, the USA for example, simply define their territories as including the marine areas over which they have been given the right to regulate and tax mineral extractive operations as provided by the Geneva Convention on the Continental Shelf of 1958. The UK however has adopted a different approach which in effect is that, in so far as operations conducted on the UK continental shelf are concerned with the extraction of oil and gas from its subsoil, they are regarded as carried on through a branch situated in the UK. In this way the majority of UK North Sea oil operations fall into the UK tax net, but there are exceptions and the precise meaning and effect of the legislation is not always clear nor certain. However, all production profits are undoubtedly caught, because licences to produce are granted to companies only if they are resident in the UK. Nevertheless the UK continental shelf remains outside the UK for tax purposes and this can be important in detailed planning.

Another important factor is the relationship between the tax imposed on profits and the tax (if any) imposed on dividends. In Germany for instance the so-called split rate system is operated, under which distributed profits are subjected to a lower rate of corporation tax than retained profits. Between 1965 and 1973 the situation in the UK was exactly the opposite. Since 1973 however, with the introduction of the imputation system into the UK, the burden of corporation tax does not depend upon the level of distributions — it is said to be neutral as far as distributions are concerned. The tax treatment of dividends for corporation taxes is important. Usually, but not always, the receipt of domestic dividends does not give rise to an additional corporation tax liability. Dividends from overseas companies are discussed in a later part of the chapter.

There are as many different sets of rules for defining profits for corporation tax purposes as there are countries imposing these taxes. It would be fruitless therefore to attempt any kind of summary, but a few important matters may be touched upon briefly.

Firstly, most countries do not have the concept which is enshrined in the UK tax law that, when certain conditions have been fulfilled, the company has commenced to trade — this is now accepted as far as production is concerned as being when a decision is made to exploit a discovery commercially. The date of commencement of trading is very important since many ordinary day-to-day expenses incurred before then do not qualify for any tax relief, and no relief can be obtained for any expenditure whatever — including that on exploration and on the many huge capital projects which are a feature of the oil industry — until the company in question has begun to trade. In most countries companies are regarded as carrying on a taxable activity as soon as they begin to operate.

Secondly, there is no feature of UK tax law which corresponds to the US percentage depletion allowance. This is a deduction from taxable profits, which is given to persons having a direct economic interest in oil in the ground to protect them against the erosion of their capital. The deduction is the going percentage depletion rate of the gross income, limited to 50 per cent of the net income. One consequence of this method of calculation is that the deductions available can and often do exceed the capital costs incurred. This contrasts with the UK system of capital allowances which restricts the deductions claimable to cost. It should be noted that although the percentage depletion allowance sometimes attracts unfavourable publicity on the grounds that it is excessively generous, it may not be as valuable to the US oil industry as the flexibility available in claiming double taxation relief.

Thirdly, some countries encourage overseas companies to invest in their territories by offering taxation incentives of all sorts, from complete exemption for a specified period, to special rules for computing income — this can be very significant in making investment decisions.

Finally, the so called 'ring fence' concept in UK taxation must be mentioned. This consists of an elaborate web of rules written into the UK tax code which, for corporation tax purposes, isolates UK and UK continental shelf production profits from other profits and losses of the same company or group of companies. The profits of the ring fence trade are calculated by reference to open market value of the oil, and losses incurred by the same company or group of companies outside the ring fence cannot be used to reduce the corporation tax on profits earned inside the ring fence, although losses incurred inside the ring fence may be set off against profits outside it. In that sense the legislation acts like a one way valve with the result that tax is paid on the full amount of the ring fence profits without any offsets or reliefs.

Double taxation and other international matters relating to trading profits

Double taxation arises potentially when two tax regimes have a legitimate right to tax the same profits. This possibility clearly exists when a company is earning profits through a branch situated in a country other than the country of residence, as will be seen from the foregoing discussion. Similar problems also arise when dividends, interest, royalties and so forth are paid across national boundaries; these are considered later in the chapter. Most tax systems have provisions designed to alleviate the worst features of such double taxation situations. Relief given in this way is usually called unilateral relief. Generally speaking, however, and certainly so far as the developed nations are concerned, relief is given mainly through the medium of double taxation agreements. These are designed to prevent double taxation and to give a measure of relief where double taxation cannot be avoided. In general a tax cannot be imposed by a double taxation agreement where no right to tax exists in the basic law, but the agreements define and regulate the granting of the various reliefs and exemptions; most importantly they define the circumstances in which a resident of one country is to be regarded as carrying on a trade in the territory of the other country, so as to become liable to taxation in that other country.

In all cases of double taxation the nub of the problem is to determine where the source of the income is situated, and who has the prior right to tax that source. In the case of commercial profits derived from such activities as production, refining and marketing, there can be little doubt that the income has its source or arises where the activity is carried out. That country, by common consent, is given the prior right to tax the profits, and the country of residence of the company will usually give credit against its domestic tax for the overseas tax already borne on the profits and will collect only the excess, if any, over that amount. Thus, if an American company is operating in the Netherlands, US tax on the Dutch profits will be exigible only to the extent that the US tax exceeds the Dutch tax on the same profits. In the opposite case the American fiscal authorities have the prior right to tax and the Dutch fiscal authorities give credit for the US tax suffered in the course of their assessment and collection procedures, so that the Dutch tax is borne only to the extent that the Dutch tax exceeds the US tax. Virtually all double taxation agreements include conditions which have to be met if a company is to be regarded as carrying on trade in the overseas country through a branch — normally called a permanent establishment. If the activities fall short of those conditions the profits are usually exempt from overseas tax. Most (but not all) oil activities meet these basic conditions.

Naturally these double taxation calculations are not as straightforward as they sound. In many cases there will be significant differences between the methods of measuring the profits in the two countries concerned. The actual tax bases may be different — one country for example may tax profits year by year as they arise, whereas another may take the previous year's profits as the measure of the income to be taxed although this is becoming increasingly rare. However in practice, the greatest differences that are likely to arise are in the treatment of such items as depreciation, drilling costs (in the case of production companies), movements into and out of reserves, royalties and interest. Differing treatment in respect of these items can have a marked effect on the quantum of double taxation relief that is actually obtained. For instance, it is possible for income as computed for tax over a period of, say, ten years to total the same in the country of residence as in the country of origin. If therefore the rate of tax in the former were 50 per cent and the latter 45 per cent, logically one would expect the fiscal authorities of the former country to collect 5 per cent and the fiscal authorities of the latter 45 per cent of that total sum of profits — a 50 per cent burden overall. If however, due to differences in the rules of ascertainment of income, a disproportionately high quantum of the income arises in the resident company's tax computations in the earlier years, and a disproportionately high quantum in the country of origin in the later years, the effect could be damaging. In the earlier years the effective overall rate of tax on a disproportionately high quantum of profit would be 50 per cent and in the later years a disproportionately high quantum would bear tax at 45 per cent, possibly giving a weighted average for the period much in excess of 50 per cent. Some tax regimes have provisions designed to iron out these sorts of inequalities.

In some countries double taxation relief may be claimed against domestic tax on a group basis — that is to say, instead of each company being dealt with individually, a group of companies is dealt with on some kind of aggregate basis. A group for this purpose usually implies a minimum percentage of common ownership of more than 50 per cent. Additionally, in the USA for example, groups may elect for treatment on a country by country basis, in which case the total tax of all the companies in the group borne in a given overseas country, may be set off against the US corporation tax attributable to profits from all group operations in that country. Alternatively a worldwide basis can be adopted, in which case the total group overseas tax is averaged over the foreign profits of all companies operating overseas, an arrangement which can be advantageous if some overseas territories have a higher tax than that in the USA and some a lower rate. Since strange and unexpected effects can be caused by the complications inherent in such sophisticated legislation, special care must be exercised in any instance where it might apply.

Another difficulty which must be recognised is that not all taxes qualify for double taxation relief. When two countries enter into a double taxation agreement, the taxes which are to be included in the arrangement are listed. They usually include additionally 'any identical or substantially similar future tax.' Where there is no double taxation agreement some countries give credit relief to their own resident taxpayers unilaterally in respect of taxes paid overseas. However, when drawing up double taxation agreements and deciding whether a new tax is 'identical' or 'substantially similar', and when giving unilateral relief, decisions have to be made as to whether a given impost is a tax or not. Production royalties are not usually regarded as taxes, and doubts have been expressed as to whether the UK PRT is necessarily a tax for double taxation purposes, because of its unusual characteristics summarised above. It is a moot point whether these doubts are likely to be resolved by its inclusion in the new Anglo/US Double Taxation Agreement since it may still be argued that it would not otherwise have qualified for relief, on the grounds that it is 'identical' or 'substantially similar' to other taxes on income. If a tax does not qualify for credit relief, it can generally be treated as an expense in the tax computation of the country of residence.

There may be particular problems associated with oil exploration and production in areas such as the North Sea where oil fields may straddle the dividing line between the marine areas of two countries. The ensuing complications may have no helpful precedent and will usually have to be solved by mutual agreement between the companies and the countries concerned. The situation of service companies — that is companies engaged in hiring rigs, barges, drilling equipment and the like to concessionnaires, and those rendering support services such as drilling, supply, diving, pipeline laying and so forth — which have their tax residence in a country other than those in which the oil reserves are situated, may prove to be peculiarly difficult in these circumstances. In all such cases it is desirable, so far as is practicable, to ensure that tax is imposed strictly within the terms of the law, since otherwise a company might find itself in the position of paying tax as a consequence of a negotiated agreement — sometimes referred to as taxation by consent or contractual tax. Such taxes may technically not be taxes at all and therefore may not qualify for double taxation relief. Similar difficulties may arise in cases where an exaction which purports to be a tax on profits is more in the nature of a royalty.

Although most cases of the potential double taxation of commercial profits are dealt with by credit relief provisions as outlined above, the profits accruing to shipping and air transport enterprises are generally dealt with by a system of exemption, under which only the operator's country of residence has the right to impose tax.

It is frequently asserted that international companies, particularly international oil companies, channel profits from high tax rate countries to low tax rate countries by the adjustment of transfer prices — that is the price at which commodities are sold across tax borders — in intra-group dealings. It is doubtful whether such practices have ever been widespread, but it is certain that the possibilities of carrying them out nowadays are severely limited; moreover attempts to do so may result in an element of double taxation for which there is no relief. Anti-evasion legislation is usually aimed at ensuring that any excess of the price of goods imported over market value is not allowed to the importer as an expense for tax, and conversely that any short fall in the sales price below market price of goods exported is added to the exporters' profits for tax purposes. In either of these events, the additional profit taxed will already have suffered tax in the other territory; in that sense there will be absolute double taxation for which no relief is claimable. Care must be taken therefore in considering such problems, to ensure that both sides of the transactions are considered. There may be cases in which artificial minimum prices are imposed upon the export of oil produced in a given country, and this may have an effect on the tax situation of the purchasing company, notably if it is a UK resident. Most double taxation agreements include clauses empowering the parties to exchange information with the object *inter alia* of aiding the enforcement of the pricing regulations.

It cannot be overemphasised that, in every instance where double taxation is involved, the interested parties should study the terms of any relevant double taxation agreement. Any reader particularly interested in the subject is also recommended to study the text of the model agreement prepared by the Organization for European Economic Development (OECD), the form of which is being followed increasingly in the negotiation of new agreements. It may also be noted that there are proposals afoot for an EEC Tribunal to be formed, which will decide in cases of dispute as to where and to what extent tax will bite when two EEC countries each claim the right to tax a given fund of profit.

Corporate structure, equity capital and dividends

It is necessary to consider some aspects of corporate structure as a prerequisite to considering the effect of taxation upon dividend flow. It is self-evident that the holding company structures of the large oil companies and the large financial and other institutions which are likely to participate in oil operations are already established and are likely to retain their present form. No one visualises Exxon, ICI or a large merchant bank, making significant changes in the top echelons of its corporate structure. Therefore, it is in setting up wholly owned subsidiaries and/or joint operations that the options become a live issue and a choice has to be made, albeit against the background of the constraints mentioned below.

The difference in tax treatment between an operation conducted through a company resident in the place where the activity is carried on, and one conducted through the branch of a company resident elsewhere, has been touched upon above in so far as trading profits are concerned. But another significant point must now be made, namely that, in the case of an operation conducted through a company resident locally, tax liability arises in the country of residence of the shareholding company only when dividends flow back to the shareholding company, whereas in a branch operation it arises in the country of residence on total profits earned (subject to double taxation relief). Dividends in the latter instance therefore are usually of little tax significance. Although the choice between using a resident company and a branch of a non-resident company is one which is becoming increasingly influenced by political considerations and is therefore effectively being taken out of the hands of the investor, it is usually an interesting and worthwhile exercise to compare the tax consequences of the two *modus operandi*. Advantages will usually be found in each method, and, even if the non-fiscal considerations preempt the decision, a knowledge of the taxation pros and cons can be a valuable adjunct in negotiations.

As an example of the potential advantage to be gained by a branch operation, one could quote a US company which might wish to operate overseas through a local branch of a US corporation in order to secure not only the US depletion allowance, but also the most advantageous group double taxation situation as outlined above. In any event, the tax rates imposed upon branch operations might be lower than those on resident company operations, as was the case in the UK until 1973, taking into account taxes on dividends. On the other hand, a local company might attract lower tax rates, or the use of such a company might be a condition precedent upon borrowing facilities being made available, or tax incentives such as pioneer industry reliefs being granted. In the final analysis, the use of a locally incorporated and/or resident company may be, and often is, a political requirement imposed for nationalistic or prestige purposes. A frequent concomitant of this obligation is that local nominees must be appointed to the board.

It is perhaps appropriate therefore to mention on the general subject of directorial appointments that it can be a mistake to allow the need to make directorial appointments — whether inspired by local or corporate requirements — to dictate corporate structures, because one can easily end up with too many or too few companies and directors. It is usually possible to devise functional titles which are just as effective for prestige purposes as directorial titles. Additionally, job responsibilities should be sufficiently clearly defined for them not to be tied to corporate structures. If it is necessary for reasons of prestige for persons to hold directorial, vice-presidential or other corporate rank, it is usually possible to set up a non-operational company for this purpose if no operational company with the necessary vacancies exists. However, in applying the above philosophy it should be kept in mind that it is often unsatisfactory for a company to carry out branch operations in more than one overseas territory, since disputes can and often do arise over the apportionment of expenditure between territories. Such disputes may easily lead to a proportion of the total expenditure not being deductible for overseas tax purposes in any of the territories, and this can prove very costly, particularly if the operation is an extensive one.

When an operation is to be conducted through a branch, the corporate structure is not likely to present many difficulties. The shares in the company concerned are likely to fall naturally into the groups shareholding pattern, and finance can usually be provided through intra-group loans which may or may not be interest free. Profits can often be repatriated by repaying loans without tax cost.

The use of a locally resident and/or incorporated company can present more difficulties, particularly if the operation is not wholly controlled by one shareholder. The ratio of loans to share capital must be decided, and this can be particularly important if the operating company itself intends to borrow in the open market. The taxation burden imposed on dividend flow can vary significantly as indicated below, and these variations must be given careful consideration.

If it is a joint venture, there is potentially a choice between using a jointly owned company and a series of companies — each a 100 per cent subsidiary of its own ultimate proprietor — acting together as a consortium, as described in the summary of UK PRT above. Such an arrangement could possibly reduce the taxes imposed on dividend flow. In this context one should not overlook the point that group accounting requirements vary from country to country, and the choice between a joint company structure and a 100 per cent subsidiary/consortium structure might also be significant in the context of the presentation of public accounts. In settling loan/share capital ratios, consideration may need to be given to legislation such as the so called 'thin capitalisation' legislation in the USA. This is designed to prevent tax avoidance by the use of high loan capital and correspondingly low share capital, as interest is generally deductible from profits for tax whereas dividends are not. It should also be noted that the UK revenue regards the restrictions imposed upon the deductibility of interest in the ring fence legislation as being tantamount to thin capitalisation legislation; however this view is not generally accepted by taxpayers.

It has been explained that the profits of a venture, carried on overseas through a locally resident company, do not fall into the domestic tax net of the shareholding company unless or until there is a dividend. Tax regimes vary in their concept of what constitutes a dividend — is it sufficient for it to be declared, or must it be remitted in cash, or is it enough for it to be credited to current account? Furthermore, are dividends taxed on the basis of the amount arising in the current year or in an earlier year? All these matters must be studied carefully.

Most tax regimes require tax to be withheld from dividends and passed over to the fiscal authorities. The rates vary significantly and are sometimes at differing rates for resident and non-resident shareholders of the same company.

If the recipient of a dividend is a resident of a country which has no double taxation agreement with the country of residence of the company paying the dividend, the recipient will be taxed solely by reference to the provisions of his domestic tax law. In some instances the domestic tax laws give relief unilaterally along the lines discussed below, but in such cases there is naturally no limit on the level of withholding tax which can be mulcted at source by the country of origin.

The source of a dividend for the purposes of double taxation agreements is generally accepted as being in the country where the company paying the dividend has its tax residence. By common consent it is also the practice to accord to that country of residence the right to withhold tax up to, say, 15 per cent of the dividend, although there may be a lower limit in the case of shareholders owning a substantial proportion (say 25 per cent or more) of a company's shares. The country of residence of the shareholder has the right to tax the recipient of the dividend and in so doing normally gives relief for the tax suffered by the source company. However the position is complicated by the existence of a number of interacting factors. Typical legislation on the double taxation of dividends recognises two sorts of tax; namely withholding tax deducted from dividends, and tax suffered on the profits out of which dividends are paid — generally called underlying tax — and two broad classes of shareholders; namely those who hold a substantial proportion of the shares in a company (direct investors) and those who do not (portfolio investors). The dividing line between direct and portfolio investment is drawn in a variety of ways, but a common test is that a direct investor is one owning more than 10 per cent of the shares and a portfolio investor less. Sometimes there are different rules governing the relief claimable by corporate shareholders and individual shareholders. Most but not all shareholders can claim credit for withholding tax, but in many instances relief for underlying tax is only available to direct investors. This is a field in which there are so many variations in detail that each case needs to be studied very carefully, paying attention to the precise wording of the law and the interacting factors which have just been considered, since any one might easily prove to be important.

Care must be taken in cases where dividends pass through an intermediate company on their way to the ultimate shareholder, since in many cases tax complications may ensue which are potentially costly. On the other hand it is sometimes possible to channel dividends through a country which has advantageous dividend withholding tax provisions written into its double taxation agreements with the paying and receiving countries; the consequence is that a greater sum after tax accrues to the ultimate shareholder than would be the case if the dividend were paid direct. In the use of such arrangements the danger of artificiality is particularly real and investors proposing to adopt such a route should prepare the ground accordingly.

Loan capital, its servicing and repayment

From the tax standpoint, and in the context of the present study, loan capital differs from share capital in two subordinate ways. Firstly, interest on loan capital is generally a deduction in calculating profits for production tax and corporation tax purposes whereas dividends are not, and secondly, loan capital, even in subsidiary or associated company activities, is often borrowed from third parties — other companies, banks, charities, institutions and so forth — in the open market, and consideration of such third parties' tax positions can be important.

It is of course incorrect to say without qualification that interest is deductible from profits for tax purposes. We have already seen that it is not a charge in arriving at assessible profits for UK PRT. In addition, many tax regimes impose conditions that must be fulfilled if a deduction is to be validated; eg the rate of interest and all the related conditions must be commercial, or excessive interest will be disallowed under thin capitalisation laws, or (for UK ring fence corporation tax purposes) the money borrowed in respect of which the interest is paid must have been used for certain specified purposes, and so forth. A great variety of conditions are imposed, but it is nevertheless a fair generalisation to say that, given the assistance of efficient tax planning, interest paid by commercial undertakings is tax deductible in most instances. Considered in the simplest light, a payment of interest merely represents a transfer of taxable income from the payer to the payee. If both are resident in the same country the effect will.be simply a corresponding transfer of the tax liability, although the mechanics of achieving this result may well be obscured by complicated withholding tax and other mechanical devices for its collection and the prevention of evasion. Nothing further needs to be said about such situations, but when interest crosses tax borders, care must be taken to ensure that the tax position both of the payer and payee is equitable.

For double taxation relief purposes there is often difficulty in determining where interest arises; therefore some double taxation agreements contain provisions covering this matter. Whether or not there is such a determination, it is customary to regard interest as being taxable both in its country of origin and in the country of residence of the recipient. In such cases, double taxation agreements usually restrict the former imposition to 15 per cent of the amount of the interest; the latter country normally gives credit for the withholding tax. The question of relief in respect of underlying tax does not arise because interest is normally a deduction in arriving at taxable profits.

Care must be taken to ensure that loans are set up in such a way that not only does the company paying the interest obtain a tax deduction, but also that the company receiving the interest has a tax liability against which any withholding tax mulcted can be set. It is in this respect that the taxation status of potential lenders must be considered. For example, if loan interest from overseas is received by a UK company under deduction of a withholding tax, no difficulty usually arises. The total tax burden of the recipient consists of withholding tax (say 15 per cent) and UK corporation tax (the balance — say 37 per cent) totalling the current UK corporation tax rate of 52 per cent. If however the loan is made by a company which has a tax loss, it will avoid the UK corporation tax liability by offsetting the income against the loss, but will be left bearing the withholding tax. Similar results can arise in respect of interest received by any entity which is not liable to tax by virtue of exemption, charitable status or any other reason. To put it another way, it must be ensured that the withholding tax is part of, and not additional to, the lenders unavoidable tax burden.

Careful consideration must be given to the currency in which a loan from abroad is to be raised and repaid, particularly if there is the possibility of exchange fluctuations which might give rise to a loss or gain on repayment. The taxability of gains and the allowance of losses against other taxable profits naturally varies from country to country. It must be self-evident however that in these times of large exchange fluctuations the sums involved can be significant. This has certainly been the case in the UK over the last decade where the possibility of introducing relief for such losses, hitherto unavailable, is being widely canvassed. If relief is given it will follow logically that gains hitherto tax-free will become taxable. It is mentioned *en passant* that, in the UK at least, eurocurrency loans do not attract any particular tax penalties or reliefs.

The flow of services between head office and subsidiaries

This can be considered broadly under two headings, service charges and royalties. The term 'service charge' is used generally to cover any intra-group charge up, down, or across the corporate structure in respect of any service rendered or facility or benefit made available which is not one of the commercial activities in which the companies concerned are normally engaged. The most common include administration charges made by head office to subsidiary companies, charges for staff seconded, or for special jobs done and so forth. The approach should be to deal with such transactions on a proper commercial basis. In all cases the most important thing is to ensure that the charges are fare and are defensible by arm's length criteria, so that the danger of unavoidable double taxation as outlined earlier is avoided. Care must also be exercised to make sure that the tax regime of the paying company treats the payment as a tax deductible expense; some regimes specifically disallow such expenses. In some cases of doubt a change of name will sometimes be acceptable; 'head office administration charge' may not be deductible but 'charge in respect of the following (listed) facilities' might be. If the overseas operation is conducted through a branch, it may prove to be particularly difficult for the branch to claim as tax deductible any head office charge, especially if it is of a vague and indeterminate nature. In any instance where it does not seem likely that the paying company will be able to claim the cost as tax deductible, it may be preferable not to make the payment at all, since failure of the payer to obtain a tax deduction will not prevent the recipient from being taxed on the income.

The term royalty is used here not in the sense of mineral extractive royalties, but in relation to copyrights, patents, designs, trade marks and the right to use secret processes and know-how. As may be expected, the relevant law is diverse and complicated. Overall however, the treatment of these royalties approximates to that of interest, with the difference that there is a general acceptance in principle that the country of residence of the grantor of the right from which the royalty flows has the right to tax the royalty. This is reflected in generally low or nil rates of withholding taxes applicable to royalties.

Miscellaneous matters

Most countries impose heavy indirect taxes on most if not all oil products consumed in their territories. These include customs duties, excise duties, sales taxes of various sorts including VAT, and turnover taxes. These taxes differ from those which have already been considered — the so-called direct taxes — in two basic ways. Firstly, they are not calculated by reference to profits but rather by the quantitative means of so much per gallon, ton, or cubic metre. Secondly, their burden falls more specifically and identifiably on the consumer than is the case with the direct taxes; indeed in the case of the sales taxes the burden only arises when the product is supplied to the consumer. The main significance of such taxes is found in the demands which they make on the working capital of the oil companies concerned; this must be taken into account when deciding how much finance will be needed for a particular project. Value added tax — a sales tax collected in a rather complicated manner — is unusual in that in certain circumstances its collection rules can actually contribute to working capital. However, this does depend upon the interrelationship between the dates of settlement with the VAT authorities, and the dates on which payment is made to creditors and money received from debtors.

Tax havens are frequently but erroneously cited as being the panacea of many tax ills. However it will be recalled that all profits earned must, of their innate nature, suffer tax in at least one place — namely the country where the activity giving rise to the profit is carried on — unless that country itself happens to be a tax haven. To that extent, therefore, there is a hard core of tax which must be paid and which cannot be avoided by the use of tax havens. Of course it may be said that the channelling of dividends and interest through a country having favourable withholding tax rates as cited earlier, makes that country a tax haven in that particular context. But other uses of tax havens are too esoteric for inclusion in this book and are, in any event, not sufficiently significant to warrant attention.

Most developed countries and some others impose tax on capital gains. Therefore the provisions of the relevant law must be watched carefully, particularly when repaying loans, altering capital structures and disposing of substantial assets, trades or parts of trades and shares. Most double taxation agreements include provisions for relief from the double taxation of capital gains.

A multitude of other minor imposts must be mentioned, including capital duties on the creation and issue of capital, loans and the like, stamp duties of all sorts, social security taxes, local and municipal taxes, wealth taxes and payroll taxes of all sorts — the list is endless. In each case it is necessary not only to establish how much is to be paid, but also whether the cost is a deductible expense in calculating any of the direct taxes discussed above.

Further Reading

For the reader who is interested in pursuing his study of taxation, the following publications will prove to be of interest: —

J D R Adams and J Whalley, *The International Taxation of Multi-national Enterprises,* (Associated Business Programmes Ltd, under the auspices of The Institute for Fiscal Studies)

Milton Grundy (editor), *Grundy's Tax Havens,* (The Bodley Head and HFL (Publishers) Ltd.)

The Board of Inland Revenue, *Income Taxes Outside the United Kingdom* (HMSO)

Hayllar and Pleasance, *UK Taxation of Offshore Oil and Gas* (Butterworth and Co. (Publishers) Ltd.)

Fiscal Committee of OECD, *Draft Double Taxation Convention on Income and Capital,* (available through HMSO)

10

Insurance of
Oil and Gas Operations

There are several reasons for insuring apart from the obvious one of recovering insured losses. With the vast capital outlay needed today to operate effectively some companies have capital requirements many times in excess of their issued capital. Loans are often necessary, and acceptable insurance with good security is a normal requirement by prudent bankers. Insurance rates are a fraction of the current interest rates. It may be said that the capital structure of an assured's enterprise will not be wiped out by an insurable loss, and his cash flow has protection. Insurance by spreading risks worldwide can even in local catastrophic circumstances protect national economies.

In dealing with such a complex subject it is not feasible to cover all aspects in detail in a short chapter, but only to give a working understanding of the way in which the business is insured and the features which are of particular concern to underwriters when assessing the risks involved.

The insurance business

There are basically three methods of recompensing an assured for losses to property and the associated liabilities following an insured peril, namely:

The commercial insurance market. This market is shared by the many individual syndicates who underwrite business at 'Lloyd's', and the national and international insurance companies. It is the prime method of insurance used by the oil industry. For many years London has been the centre of the world's international commercial insurance market.

A mutual association. Essentially this involves a group of individual firms with like interests entering into a funding arrangement whereby an initial call, or premium, is paid according to

Note: The terms of the insurance policies quoted in this chapter are those prevailing in 1977. All values are expressed in American dollars.

entered assets. Any losses covered under the terms of the arrangement will be shared between the participants. The narrow base of such an arrangement is the essential weakness, for several large losses within a short period of time will make it an expensive 'club'.

Self-insurance. This is self explanatory, however complete self-insurance is unusual (except for governments) and companies who do decide on this method may purchase catastrophe insurance from the commercial market.

The petroleum industry is a very complex and often hazardous class of insurance business with tremendous values exposed both for physical damage and potential liabilities. The London market provides the highest capacity available to the international assured for the very high values involved in insuring, with good security, many of today's insurances or 'risks'.

The underwriters or risk takers

The world's insurance markets normally classify business as 'marine' or 'non-marine'. Separate companies, divisions within companies, or syndicates in the case of Lloyd's, employ underwriting personnel who specialise in marine or non-marine operations. Some aspects of the petroleum industry entail a mixture of marine and non-marine risk exposures.

Each Lloyd's or company underwriter has a limited capacity for his acceptance of a line, or share, in any one risk, always assuming that he considers the risk is one that he is prepared to underwrite. For small values or amounts the underwriter may write a substantial percentage, but as the sum insured rises his share must prudently be reduced — he must not endanger his reserves on any one risk or indeed on an accumulation of risks in any location.

If an underwriter decides the petroleum industry is a class of business he is prepared to underwrite then he must decide on his underwriting policy. With any complex class of business certain underwriters specialise, and are accepted by their fellows as 'leaders'. Thus with oil operations business there are recognised leaders in the insurance market.

The underwriters, or risk takers, would not be able to function effectively without the other members of the 'team'. There are specialist surveyors and some insurance companies have 'in house' surveyors, or may employ independent experts, and organisations like the London or United States Salvage Associations, and the classification societies (such as Lloyd's Register of Shipping) are often employed. The claims adjusters, lawyers and many others also ensure that a great fund of knowledge and expertise is available to meet or advise on every eventuality.

The role of insurance brokers

To bring an insurance proposition to underwriters a qualified broker is normally employed, for an underwriter at Lloyd's can only accept business via a Lloyd's broking firm. To be a Lloyd's broker means that the Committee of Lloyd's have accepted the firm's credentials. Certain brokers specialise in oil operations business. In the London market most of the insurance company dealings are also with Lloyd's brokers, for the broker must frequently approach many Lloyd's and company underwriters to complete a risk. The majority of highly valued risks are also subject to international competition. The jumbo-sized risk may require the participation of many of the world's insurance markets, either direct or by way of reinsurance.

The broker, it must be understood, is the agent of the assured, not of the underwriters. When engaged by an assured he has to prepare his 'brief' for the underwriters. This may involve him in a great deal of expensive correspondence and research. The first underwriter he sees will generally be a specialist leader in the type of risk that he has to place. The broker often goes to several specialist leaders before finalising the details of the insurance that he has to place for the assured. He is, of course, obtaining the best terms possible for his client before approaching other underwriters in the market for their support in order to complete the risk.

The qualified international insurance brokers often work with local national brokers or business producers with specialist knowledge.

Premiums, claims adjusting etc.

The surveyor's report will be the major tool used by the insurers to calculate the premium and the share of the total insurance that each underwriter will be prepared to accept. The surveyor puts a figure on the 'estimated maximum loss' (EML) in the event of a serious fire or explosion. It follows that the higher the EML the lower the percentage each insurer is prepared to accept.

To determine a premium equitable to all parties the underwriter examines the fire and explosion hazards, the values at risk, together with all the other relevant information, particularly the surveyor's reports. To arrive at an average rate per cent, differential rates are applied to the various components of the risk. In calculating the rate or premium for the insurance, underwriters take into account the loss experience of the assured and also their experience in insuring similar plants for other assureds worldwide (plus his knowledge of uninsured losses).

The deductible is negotiable, and it is up to the assured to decide what he is prepared to accept, according to the price quoted by the insurers for, say, $100,000 or $1,000,000 deductible.

When a casualty occurs which may lead to a claim under the

policy the claims department of the leading underwriter is immediately informed by the assured directly or via his broker. The adjustment of claims is a specialised profession and the action taken will depend on the particular circumstances of the incident in question.

The claims adjuster generally requires an independent survey report of the damage, and competitive tenders for repairs. Any remedial action to be taken, depending on the severity of the damage, will normally require acceptance by the underwriters' surveyor.

Casualties often involve a complex set of circumstances with costs and expenses (apart from the deductible or any self-insured amount) being partly for the account of the assured and partly covered by the insurance. Hundreds of individual invoices may have to be analysed and correctly allocated in the event of a severe casualty.

Independent claims adjusting organisations or average adjusters are often employed by the assured to marshall the facts and figures for presentation to the claims department of their insurers.

Kinds of cover

The kinds of cover provided are limitless; but in respect of the petroleum industry they can be conveniently considered under four headings:
—the construction phase
—the testing and commissioning phase
—the operational phase
—special risks such as earthquakes, wars, business interruption etc.
It can be seen that insurance is available from the 'cradle to the grave'.

Construction and commissioning phases
The starting point for the assessment of risks is the design of the unit and the capability of the firm undertaking the construction. Whether considering a drilling rig mounted on a land vehicle, an offshore platform, a drill ship, or a refinery complex, the technical expertise of the designers and the building firms doing the job are of prime importance — human error being still the major cause of most accidents.

However there is an element of trial and error in all construction and especially marine construction, and until a design has been tested fully in all types of conditions and any problems arising solved, the structural strength and seaworthiness of a newly designed unit must be treated with caution.

In making these assessments underwriters do not pretend to be engineers so they must rely on professional advice, and the technical appraisal is obtained from classification societies, and the specialist surveyors. These surveyors may be employees of an insurance company or be members of a separate specialist survey firm.

Consequently with the more complex designs, despite the reputation of the designers and of the building firm or consortium, the approval of independent experts will be sought at the discretion of the insurers. This applies particularly to designs for offshore structures where local weather and seabed conditions are extremely important factors to consider. The North Sea and adjacent areas have probably the worst weather of any of the offshore oil drilling areas of the world. There are stronger currents (for example in Cook Inlet in Alaska) but the seabed being generally sandy is constantly subject to remoulding by the strong tidal currents and storms, and the weather is extremely changeable in the North Sea — dense fog, strong winds, snow and 'black ice' are common in winter. Special premium loadings have been applied for northern North Sea operations. It is true that the major hurricanes of the US Gulf are more violent, but the consistent year round adverse weather pattern is absent.

The underwriters and technical experts must consider carefully the testing and commissioning phase of each project, concentrating particularly on the breakdown and explosion risks (compressors and boilers particularly are high risk items). Damage due to faulty or defective design is a serious risk and care must be exercised in selecting risks for insurance. The costs incurred in rectifying the design defect are not covered. Exclusion of certain specific parts of the plant may be required and high deductibles, especially during testing, are mandatory.

Construction risk insurance is available in the commercial market, and included under this heading will be the trials, testing and commissioning risks. Policies are issued for the period of the construction risk until final delivery to the owner. This period is stated so that any delay in delivery would call for a reassessment of the premium.

Operational phase
At the operational stage an additional group of factors have to be taken into account depending on the nature of the unit and the operations involved as will be discussed in subsequent parts of this chapter. Take for instance the movement of marine drilling rigs from one location to another. Where towage is required first class tugs of sufficient power must be available and used. The operation should be properly supervised by a surveyor approved by the underwriters. The weather conditions, unless in the case of an emergency tow, have to

be monitored for a satisfactory forecast. In the case of a long trans-ocean tow, the weather cannot always be predicted far in advance, but if possible a good weather season should be aimed for. The premium charged for such tows will naturally be variable according to the type of unit concerned, the length of tow, the likely weather conditions and the season and the reputation of the towing company and tug, not forgetting the experience of the crew.

Good management is absolutely essential in any enterprise. Safe operating procedures, a good recruitment and training programme, good staff relations with a contented (but not self-satisfied!) staff, and other management factors, obviously pay dividends in efficiency and production so that, in the long run, fewer claims from the insurance view point can be expected.

Fire, lightning and explosion risks are the minimum physical damage coverage normally required and, possibly depending on the geographical location, additional perils e.g. wind storm, flood and earthquake, may be required.

There is also a specialist market providing coverage against general liabilities and umbrella type coverage, substantial deductible and/or excess points are applied and necessarily each case has to be examined and underwritten separately. Primary employer's liability and property and product's liability coverage is available mainly in the non-marine market.

Loss of profits or business interruption

Loss of profits or business interruption can be insured. The factors that the underwriter takes into account for the physical damage insurance will similarly be of considerable importance for this type of cover.

In so far as a physical damage loss probably will lead to a loss of profits to the assured, the underwriter is faced with an increased loss potential and the capacity of the market will be fully tested by the largest risks. Particularly relevant to this form of insurance will be the inter-dependency of production units, and the time it takes to replace or rebuild a damaged or destroyed unit. An excess period of say 10 or 14 days, which is at the risk of the assured, is usually required. This period and the premium will be variable and depend on the type of plant, its capacity and the operating and loss record of the plant or similar types of plant in the case of a new plant.

It should be remembered that a small physical damage loss can result in a considerable loss of profit. If a large assured decided to self-insure the physical damage risk and insure the loss of profits in the commercial market the underwriter would expect to receive a higher premium level on the loss of profits insurance. Apart from selecting against the insurer the bulk premium level and spread of overall risk is reduced.

Earthquakes and war risks

One important peril, earthquake damage, is not covered in some areas of the world where earthquakes are inevitable, heavy and frequent. It is evident that the effect on the equipment concentrated over an oil field could be catastrophic. However for floating units the exclusion of earthquake may be deleted. There is a market that specialises in earthquake risks, so that cover may be purchased as a separate insurance for the hazardous areas.

Due to the massive potential concentration and political nature of war risks there is a 'waterborne agreement' whereby the commercial market will only accept war perils on waterborne units. Thus a fixed platform or refinery would not be covered for war risks.

War risk policies exclude the hostile detonation of any weapon of war employing atomic or nuclear fission or fusion or the like. War between the great powers — USA, UK, France, USSR or China — is not covered, nor is capture, seizure etc. by the government of the country in which the unit is owned or registered. Operations in certain politically hostile areas of the world are excluded from coverage, by warranty, these may vary according to political circumstances, and there is a special notice clause for underwriters to amend these warranties and the premiums.

These policies are issued for periods of not exceeding twelve months. The terms of the annual renewals are varied in the light of experience, both particular to the ownership, and general to the class. Not all underwriters are prepared to accept war risks, however the majority will accept short term voyage policies for marine and war risks combined.

Insurance at the exploration stage

Once the geologists and geophysicists have located an area of oil or gas-bearing potential, the crucial exploratory drilling is undertaken to discover whether oil or gas is present and if available in commercially viable quantities.

The first successful oil well was drilled in Pennsylvania in 1859. The first rotary system used in a major completion was in Texas in 1901. On shore, rotary drilling has been practised for over a hundred years. The first offshore operations really began from a fixed platform with a producing well in shallow waters in the Gulf of Mexico in 1938, and giant strides have been made since then. The enormous expansion of the need for oil, gas and oil products meant that land operations in the possible operating areas in the world which were known then could not supply current and future requirements. Thus increasing emphasis has been given to offshore exploration and production. Generally, whether on land or offshore, some considerations are common.

Land based rigs

The steel scaffolding, known as the derrick, which is used to suspend the drill string and hoisting gear, can be erected and dismantled piece by piece, but the portable types can be raised or lowered and moved on transporters to new locations. The size of the derrick will vary with the depth of drilling required. There are various other items of essential and complex equipment.

For the insurance of a land based drilling rig the underwriters will principally want to know the make, where and when manufactured, designed drilling depth, the purchase cost, the present estimated value, the type of rig and proposed use (whether exploratory or workover), whether directional drilling is anticipated, if any medium other than drilling mud is to be used, the intended area of use, and whether earthquake cover is required, details of loss experience, if any, for say 5 years (on any rigs operated) and how long the assured has been in business. A surveyor's report is generally called for, and an adequate valuation (based on replacement cost) will be required. Workover rigs are generally moved more frequently involving erection, dismantling and transit.

There is a standard insurance form used in London known as the Oil and Gas Well Drilling Tools Floater Form (Land) with a mandatory deductible. The policy insures against direct loss or damage by specified perils e.g. fire, lightning, tornado, cyclone, windstorm, hail, flood, blowout and cratering, raising, lowering, collapse and/or pull in of derrick or mast, etc. There are specific exclusions (wear and tear, property installed on vessels, drilling barges, piers, piling structures, earthquake risks, war risks, etc.). There is a blowout preventor warranty and a 100 per cent co-insurance clause, designed to ensure proper valuation for insurance.

Marine drilling rigs

Marine drilling involves a floating or leg supported 'platform' which can enable the derrick and drilling equipment to operate — suitably modified for the marine environment.

The shallow sheltered waters of coastal areas were the first exploration target. The prime requirement was for a stable platform on which to erect rotary rigs. Initially two types of platform were developed — one which could be compared to a grounded barge, and the other which floated whilst providing a stable drilling platform. As drilling in deeper water becomes necessary legs and a jacking system were installed on some types of platform. At this stage drilling operations had to be extended into deeper and relatively unsheltered waters; this was a new challenging and unknown frontier and many different designs were tried, and there were some stability problems and casualties occurred.

There are basically three main types of offshore units used for oil or gas drilling. A jack-up unit is a self-elevating platform which floats freely with its legs retracted by a jacking system for when it is being towed; sometimes ocean tows necessitate the temporary shortening of the legs for stability reasons. There may be three or more legs depending on the particular design. When drilling, the rig is jacked up to provide a sufficient air gap between the base of the platform and the surface of the sea to satisfy design requirements. Key considerations in determining risks are the design and efficiency of the jacking system (and the proper operation of it), the nature of the seabed (established by a survey) and the method of coping with the penetration by the legs in soft or uneven seabed areas. The majority of the many accidents that have occurred to this type of rig has been during tows, or whilst jacking up or down.

The semi-submersible or column stabilised drilling platforms can be bottom supported or free floating. Some types have no means of propulsion, whilst others have been equipped with propulsion or thruster assistance, or have full propulsion systems and carry qualified navigating crews. In general the semi-submersible units are larger and more expensive than the jack-up type. They can operate in deeper water, and usually have a multi-anchoring system employing six to ten separate lines. The stress on each cable or chain is usually governed by computer to counteract the effects of weather stress and currents in order to maintain the unit at the required location on the seabed. The major problem area is stability under critical conditions when in tow, when ballasting or deballasting, especially if the self-propulsion system has insufficient power.

The ship-shaped unit is basically a conventional ship with a slot for the drill located approximately in the centre. It normally employs a multi-anchoring or stabilising system. A full propulsion system is installed and a navigating crew is carried on board at all times. Some are custom built and others are converted merchant ships. This type of rig is capable of drilling in much deeper water than the jack-up or semi-submersible rigs.

There is also the catamaran or twin hulled type. This has to be so constructed to resist the forces of the water between the hulls, as well as overcoming the structural problem of joining the twin hulls.

The construction of marine drilling rigs such as drill-ships, semi-submersible and jack-up units, which are capable of floating and moving from station to station, either under their own power or in tow, are essentially marine risks. They are insured under an 'all risks' form of policy with certain exclusions, and the inclusion of such liabilities as damage done to other vessels or objects but limited to the value of the unit insured. The normal policy form is the Institute Clauses for Builders Risks, and deductibles vary according to the value of the unit. The premium is assessed on the final completed value, the location and type of trials necessary, and the distance of the voyage under own power or in tow in delivering unit.

For the operating insurance of the marine based drilling rig unit the underwriters adopt a philosophy similar to other marine hull business, with the necessary extensions and limitations of coverage peculiar to oil drilling operations.

Jack-up and semi-submersible units (except those which are fully self-propelled) are covered for marine risks under the London Standard Drilling Barge Form. This is basically a physical damage all risks and collision lialbility form with specified exclusions — within a defined operating area. The minimum deductible for all classes for each occurrence, except total loss, is one per cent of the insured value, with a minimum of $50,000 and a maximum of $200,000. Higher deductibles can be negotiated. A further requirement is for a 100 per cent co-insurance clause at an agreed value of the unit insured.

Ship-shaped units or drilling ships and fully self-propelled units are comparable in many respects to merchant ships and are therefore insured under the standard insurance clauses for ships (Institute Time Clauses — Hulls or American Institute Hull Clauses). To provide for the special perils allied to drilling, additional clauses extracted from the barge form are also used. Wider navigational limits, due to the independent freedom of movement of these units, are accepted using 'Institute Warranties' which restricts navigation without special agreement only in respect of particularly hazardous areas, such as the Arctic and Antarctic. An approved classification society class is normally required for these units and must be maintained. The minimum deductible is one per cent of the insured value with a minimum of $25,000 and a maximum of $100,000, but a higher deductible can be negotiated.

Loss of use of rig and other risks

In the event of a serious accident to a marine drilling unit requiring it to be towed for a considerable distance to a suitable dock for repair, it can be seen that there could well be a substantial loss of hire claim to the operator. This kind of loss of hire can be insured provided it results from an insured peril, but these policies normally exclude any claims following total loss of the unit. The usual basis is for coverage, within a 12 months policy, for say 90 days in excess of 30 days of any one occurrence. The amount per day would be an agreed sum and the premium charged at a rate per cent on the maximum indemnity.

Turning briefly to the other major marine liabilities (excluding collision up to the insured value), protection and indemnity (P & I) risks are traditionally covered by the P & I Clubs. Certain of these

Clubs do specialise in covering up to stated limits the P & I limits applicable to semi-submersible and ship-shaped units. These policies cover, in essence, the liabilities of the owners of the units, but do not extend to the risks and liabilities of the oil companies. Terms are limited in some respects, thus, for example, the risks of 'loss of hole' and pollution from the well would not be covered by these policies. Coverage of marine pollution risks is covered in the next section.

The production of petroleum

Once oil and gas is discovered in commercial quantities the drilling rig will be removed and the appraisal drilling and production platforms and pipelines will be installed. In some instances the mobile drilling rigs can be used as temporary production platforms or gathering stations.

Fixed structures
Offshore production platforms usually entail construction in steel or concrete, in many respects similar to building on land. Exceptions to this type of construction were the platforms used at Cook Inlet, Alaska. These had single legs constructed from special steel, the design was considered the most effective to resist the strong currents and the heavy ice pressures in winter.

Construction normally commences in specially prepared 'basins' on land until the platform is sufficiently completed so as to allow floating out and further construction afloat.

Platforms have usually to be assembled at sea with the aid of specialist heavy lift marine barges and ship or barge transporters. Heavy claims have resulted from the failure or inefficient manoeuvring of the cranes during the construction stage with the result that it is now recommended that such operations should be performed in daylight hours only.

It can be seen that, being permanently fixed, in most cases in salt water, the problems of marine growth and corrosion are increased, and as many wells terminate at the platform the increased hazard is self-evident.

The construction of platforms for offshore operations involves many contractors and sub-contractors and normally the complete project is insured by the oil company for whom the platform is being built and the policy will also cover the contractors and sub-contractors in all aspects of the project. This avoids problems of coverage by different policies, and the possibility of disputes between the parties and their underwriters over which party should respond in the event of a claim. Special 'all risks' policies for physical damage have been developed for these platforms, but are subject to certain exclusions and have substantial deductibles. They also cover third party liabilities. The insurance for physical damage and liability risks in respect of the 'basins' and associated workshops, plant and equipment, can be insured in the non-marine market.

The next stage is a hazardous tow to the site where the unit is sunk at the pre-determined and prepared location followed by settling or piling in and final completion on site using special heavy lift crane barges for the modules and other equipment. Delays at this final stage are common due to bad weather. The high risk marine content of these final construction stages places them within the marine underwriters sphere.

These platforms are the most highly valued and concentrated risks in a marine underwriter's portfolio; for instance the risks include the sheer size of these units — some are 300,000 tons or more — the requirement for them to operate in some 500 feet of water in the notorious North Sea weather conditions, the many engineering sciences that are working to the limits of technology, the hazardous towage and settling down and the heavy lift problems of completion on site from the crane barges. Several severe accidents have occurred — the most expensive so far being the sinking of the Frigg Field DPI jacket.

The policies for fixed platform wordings are similar to the London Standard Drilling Barge Form. However, as the platforms are fixed there is no collision clause requirement and a territorial limit is stated instead of a navigational limit. With the exception of the super platform, deductibles are a minimum of three per cent of the value with a minimum $10,000 and a maximum $25,000 for each and every loss, excluding total loss — but a higher deductible can be negotiated. Operations in areas of relatively increased hazard are subject to loaded premiums with a further loading if platforms are involved in servicing operations. Furthermore, if named windstorm and/or hurricane coverage is required as in the Gulf of Mexido, higher premiums and a special deductible apply.

Well control

Oil or gas pressures vary between the fields and between different areas of the world and naturally have an important bearing on the possibility of blowout with the attendant risk of fire and of cratering.

The cost of control insurance is, broadly, to reimburse the assured for expenses incurred to regain control of a well or wells in the event of uncontrolled blowout and/or cratering. Fire-fighting specialists with special equipment and techniques may be required, and, for a massive offshore operation, a suitable drill barge to drill a relief well may have to be hired at high cost — and possibly brought to the scene of the incident from thousands of miles away (this operation is excluded from normal coverage). The annual premium charge is greater for offshore operations and varies according to area and the depth drilled, thus the Arctic and Antarctic would have loaded premiums (sub-sea completions have an additional loading). There is a minimum deductible of $25,000 or $50,000 depending on area and a 15 per cent or 20 per cent co-insurance clause (i.e. this amount is uninsured). An extension to the coverage is available for the cost of clean-up and containment arising from any occurrence covered by the cost of control section, but the deductible is increased. An assured may elect to insure all wells for which he is liable offshore (except in the Arctic or Antarctic) or all wells onshore (except for the Arctic or Antarctic) or in all three areas. Policy limits and deductibles are on 100 per cent basis, as a number of companies, besides the assured, may have an interest in a well. With multi-well platforms, and separate wells in close proximity to one another, the policy limit is stated for any one occurrence.

Marine pollution

The great expansion of offshore oil exploration and production increased the potential oil pollution risks to the environment following an escape or discharge of oil from offshore operations.

Marine pollution following a blow-out or accident can be insured for the cost of removing, nullifying or cleaning up seepage, pollution by contaminating substances emanating from oil or gas in respect of stated operations, including legal liability for damages for bodily injury and loss of use of property. The limit is in respect of any one claim or series of claims arising out of one event, and in the aggregate, during the policy period. The assured is required to bear a substantial first loss retention. The main exclusions from coverage are fines or penalties, including punitive or exemplary damages; loss in respect of the assured's own property or property in his care, custody or control; the cost of controlling a well or of drilling relief wells; war and associated perils; non-compliance with any government rule, regulation or law applicable; or claims arising out of the transportation of oil or similar substance by watercraft. It is a warranty that the assured uses every endeavour to ensure that they and their contractors comply with all regulations and requirements in respect of fitting blow-out preventors, storm chokes and other equipment in order to minimise damage or pollution and that, in the event of blow-out or escape of oil or gas, the assured will use every endeavour to control the well or stop the escape.

Offshore Pollution Liability Agreement

A voluntary scheme, now including more than thirty international oil operating companies was initiated in 1972 by the United Kingdom

Offshore Operators' Association, to provide adequate coverage until a regional, inter-governmental convention was implemented. The British Government were fully involved and kept their European counterparts informed.

In 1974, OPOL (the Offshore Pollution Liability Agreement) was set up, becoming effective in 1975, with compensation up to $16 million per incident available to deal with oil pollution arising from offshore oil operations. The draft, applicable to UK waters, was extended to Denmark, West Germany, France, Ireland, the Netherlands and Norway and may be further extended.

The agreement provided for orderly and speedy settlement of claims, encouraged immediate remedial action, ensured the financial ability of participants to meet their obligations, guaranteed that claims would be met, and was designed to avoid complicated juris-dictional problems. Acceptance of liability under OPOL cannot supplant legal liability, but provides a simpler and more satisfactory means of negotiating claims for the claimant and the operator.

Operators, rather than licensees, participate in OPOL being more directly involved and better placed than non-operators to assume the obligations.

Within the terms of OPOL members accept strict liability unless they can establish that the incident resulted from war, or exceptional natural phenomenon, act or omission of a third party with intent to cause damage, negligence or wrongful act of a governmental authority, or an act or omission done with intent to cause damage by a claimant, or from the negligence of that claimant.

Members must provide evidence of financial ability to meet claims of up to $32 million annually, i.e. two maximum losses. Claims are jointly guaranteed in the event of default by a member. The agreement is administered by the Offshore Pollution Liability Association Ltd. Insurance is available in the commercial insurance market to cover OPOL liabilities subject to a substantial deductible for any one accident or occurrence.

Carriage of oil and gas by sea

Nearly 60 per cent of the world's ocean going tonnage is employed in carrying oil or gas products. In 1976 of some 4,000 vessels of 357 million tons dwt independent owners (as opposed to oil companies) controlled some 2,400 vessels of 235 million tons.

There are many different types and sizes of vessel depending on trade and operational considerations. At the top end of the range are the new generation of LNG tankers with insured values of $175 million. There are also large numbers of specialist support vessels servicing the oil industry such as crane barges, pipelaying barges, tugs, supply vessels, anchor handling vessels, crewboats, submarines, etc. Aviation, too, plays a vital part mainly with helicopters.

The majority of ships are insured in the commercial insurance market for marine and war perils (including collision liabilities) as discussed previously in chapter 6. There are a number of variations to the standard policy form giving different degrees of marine coverage, a deductible may be for all claims on each accident or occurrence excluding total loss. The deductible may be a few thousand dollars or, on the larger vessels, several million dollars. The premium charged depends on many factors besides the deductible but the owners background and insurance record, the age, condition, size and value of the ship and nature of the trade involved are extremely important.

The oil or gas itself, the drilling mud, pipes and general supplies are normally insured separately for marine and war perils and certain other liabilities.

General liabilities associated with the ships, their crews and cargoes are normally insured in the mutual protection and indemnity clubs, however oil pollution liabilities are dealt with separately.

Oil pollution liabilities

A particular liability associated with the carriage of oil by sea is pollution liability. Few will forget the first serious oil pollution incident following the grounding of the *Torrey Canyon* in April 1967, and the pollution of the English and French beaches when the major part of 117,000 tons of oil cargo was released into the sea

To fill a gap in maritime law TOVALOP (Tanker Owners' Voluntary Agreement Concerning Liability for Oil Pollution) was started by seven major oil companies in 1969 and over 99 per cent of the world's tankers now participate in this agreement.

The agreement establishes which court of law would have jurisdiction and the owner accepting responsibility for preventative measures or for clean up costs, unless he can prove he has not been at fault. Following an oil spill the owner either removes it himself or reimburses the government, or associated public authority, concerned for the clean-up costs reasonably incurred without waiting to be sued. Liability is up to $100 per gross registered ton or $10 million for each tanker involved in any one accident. The owner covers his liability by entry in a P & I Club or by other insurance acceptable to the International Tanker Owners' Pollution Federation Ltd. which is the administrative body of the agreement. The P & I Clubs are substantially reinsured in the commercial insurance market.

To supplement TOVALOP an oil cargo owners' scheme was established — known as CRISTAL (Contract Regarding an Interim Supplement to Tanker Liability for Oil Pollution). Its rules state that the tanker carrying the oil cargo must be a member of TOVALOP and the owner of the oil cargo must be a member of CRISTAL. CRISTAL is designed to provide compensation after other sources have been exhausted. The amount payable for any one incident is increased up to a limit of $30 million by paying the difference between the combined TOVALOP and legal liabilities and the actual loss or expense. Claims settled by CRISTAL are paid from money subscribed by its members who contribute the necessary funds in proportion to their share of the total membership's worldwide shipments. The collection and claims settling agency is The Oil Companies Institute for Marine Pollution Compensation Ltd.

Thus TOVALOP reimburses governments whereas CRISTAL will additionally pay compensation to individuals e.g. fishermen, boat owners, and anyone suffering damage from oil pollution. Both TOVALOP and CRISTAL were designed to terminate if and when an international compensation fund came into force.

The CLC (International Convention on Civil Liability for Oil Pollution Damage 1969) came into force in 1975, following ratification by the required number of states. The convention does not cover all aspects of oil pollution in those countries where it has become effective. The principal coverage is for spills of persistent oils and not for instance a fuel spill from a ship in ballast. Liability is strict, rather than based on fault; the only defences available are act of God, act of war, damage caused wholly by a third party with intent, and sole negligence on the part of a government or other authority in the upkeep of lights or other navigational aids. Liability is imposed only on the registered owner of the tanker (and not, for example, on a bareboat charterer). The convention limit of liability is $160 per limitation ton, with a maximum of $16,800,000, and applies to all claims arising out of one incident, including the reasonable costs of clean up which an owner himself incurs.

There are therefore areas where TOVALOP may have a part to play without overlapping the convention provisions. Both TOVALOP and CRISTAL are voluntary measures and both have already paid substantial sums that would not otherwise have been available, to settle pollution damage claims.

Pipelines and processing plants

Once the oil or gas has been delivered by tanker or pipeline to the refineries insurance cover is provided by non-marine underwriters. The major cover required will be for material damage and loss of profits following fire and/or explosion.

Pipelines

Pipelines are often insured for construction risks as well as operating risks after construction. There is a specialist market for this type of cover and each case has to be examined individually. Among factors to be borne in mind are type of pipe, method of laying, location (including the proximity of buildings, etc. on land and sea traffic offshore), weather conditions, depth of water, type of seabed, whether the line is buried and how deeply, the experience of the contractor and the type of equipment to be used. Depending on the value, there are very substantial deductibles. Over the years there have been some very heavy insured losses in this sphere.

Refineries and processing plants

Enormous values may be involved in complexes consisting of several oil refineries and petrochemical plants. Total values in excess of $500 million are not unusual. High deductibles are normally required; for instance, for fire or explosion $250,000. A storm or flood extension of $50,000 is not uncommon.

The insurance regulations in each of the countries concerned will have to be examined to ensure that there are no restrictive regulations applying to insurance.

The construction of a refinery complex is subject to many perils such as fire, storm, flood, earthquake and impact. For instance, the greatest risk to which partly completed tanks are subjected during erection, is adverse weather conditions. Insurance experience during this phase has been unfavourable. Most tanks because of their size and purpose are built on flat, exposed sites, often near to tanker berths. Sometimes tanks have been built on reclaimed land which may be vulnerable to flooding or shifting foundations. Considering the typical range of equipment to be installed in addition to tanks including furnaces, heat exchangers, steam turbines, distillation columns, boilers, pumps, etc. — there are many problems for the insurers, not the least being breakdown claims during initial testing.

In recent years commercial and economic considerations have tended to increase the degree of integration and capacity of this type of plant. Thus in one location there may be a concentration of equipment, minimum spacing and thus a high accumulation of values. The insurers will therefore require a breakdown of the insurance values by location showing the individual plant and tank values at each location. The insurers will also require an inspection of each location by fully qualified fire surveyors and engineers who will concentrate primarily on plant spacings, unusual or prototype processes, fire prevention and protection standards and also concentration of values.

In the case of a prototype plant the underwriter pays particular attention to the surveyor's report and evaluation. Similar types of processes or plants will be a base for the underwriter's calculation of an adequate premium.

Superior loss prevention measures by an assured may well affect the terms offered by the underwriter, particularly on renewal when proved to be effective. Thus the distance of a new unit from existing units may have a bearing on loss prevention and the 'estimated maximum loss' and thus the share of the total risk that the underwriter is prepared to accept.

Further Reading

Victor Dover, *A Handbook of Marine Insurance,* (Witherby)
Smith and Francis, *Fire Insurance Theory and Practice,* (Stone & Cox)
Oil and Drilling Techniques (Petroleum Information Bureau) — a number of other useful booklets are also available from the same bureau.

PART E
SOURCES OF FINANCE

The sources of finance available to companies in the international petroleum industry can, perhaps simplistically, be classified into the following broad categories:
- —internally generated funds by individual companies
- —government subsidised finance
- —conventional bank lending
- —eurocurrency markets
- —stock markets

The sources of internally generated finance can be subdivided into two categories: disposal of existing assets, and retention of earnings from current operations. The first approach may be either by outright sale or some form of 'farm-out' arrangement, and is characteristic of the independent oil company; it has therefore been mainly dealt with in chapter 4. The retention of earnings in the business is a particular feature of the large integrated company, and this is discussed in considerable detail in chapter 3.

Government subsidised finance may be made available for domestic use or to finance exports. Examples of the first are the credit facilities made available to shipowners by their governments for new building contracts placed with their domestic shipyards. Another example is the 'interest relief scheme' available from the UK Government for financing the manufacture of equipment in the UK for use in the British sector of the North Sea. However, the most important sources of government subsidised finance are the export credit schemes available from most industrialised countries. Generalised reference to these has already been made in the four chapters of part C. Bearing in mind its importance in project financing, chapter 11 describes the two elements of all export credit schemes, namely the privileged credit terms and the coverage of the commercial, political and other risks. It deals both with the general philosophy of this source of finance and the constraints imposed by governments and by international coordinating organisations such as the EEC and the OECD etc. It emphasises the continuously changing nature of the regulations governing this source of finance.

In the discussion on commercial loans in part C particular

attention has been paid to the approach adopted by the lending institutions to the assessment of risk and to the various ways of achieving the security that they require. These discussions are equally applicable to loans extended by domestic banks and by overseas banks and through the medium of the eurocurrency markets. However, the latter have a number of unusual features and are therefore discussed in detail in chapter 13. Eurocurrency markets can be considered as being two markets — eurocredits and eurobonds. The former are akin to conventional bank loans whilst the latter are medium to long term loan stocks. This chapter is intended to acquaint the reader with the mechanism of the eurocurrency markets and the requirements which borrowers have to meet.

The significance of equity in the financing of new projects and developments has been emphasised in many previous chapters most notably chapters 6 and 7. In chapter 12, selected aspects of the stock markets in various countries are dealt with. To be able to use these markets effectively it is essential to recognise that whilst all markets are similar in nature, there are important differences, particularly in terms of the capacity of markets, requirements for listing, methods of operation and the means of regulating the markets. It is not uncommon for eurobonds to be listed on one or more stock exchanges. Such listed bonds would not necessarily be traded exclusively through the organisations directly affiliated with that stock exchange. The principal reason for seeking a listing for an issue is that by so doing the securities become eligible to be subscribed by certain institutional investors which are required to invest only in listed securities.

This brief note highlights the technical nature and interrelationship of many sources of finance. It is beyond the scope of this book to do more than alert the reader to some of the more important areas. References for further reading are given at the end of each chapter.

11

Export Credit Finance

Shortly before the Second World War, most of the Western European countries had to develop their industrial activities in order to survive and provide employment for the millions of workers who had been directly affected by the 1929 economic crisis. Obviously, such a call for industrial expansion and development was a call for finance. Furthermore, as the home markets were still mainly in recession, each country had to look beyond its own boundaries and concentrate upon exports as much as possible in order to revive its economy by finding new trade outlets abroad.

Exporting then became the general trend. A major problem however had to be faced — recourse against the defaulting foreign debtor. Recovering one's money from a slow paying debtor has always been a major problem; it is that much worse when the debtor is in a foreign country and operating under foreign legislation. The first and easiest reaction was to avoid dealing with foreign companies, but the second and prevailing one was for governments to realise that exporting was essential to their economies and it would be of economic benefit if they provided some form of government financial support to aid export development.

In line with this concept, the United Kingdom was the first country to set up a governmental agency guaranteeing the UK exporter against defaults by a foreign importer in the reimbursement of his UK credit. France soon followed.

When the Second World War ended, the world was left in a turmoil. This economic predicament favoured the development of insurance of exports. As more and more countries became politically independent, so there was a greater need for new domestic industries to be established. This could only be made possible with the assistance of the industrialised countries.

At that time, a further incentive for export credit insurance was that not only was there the straight commercial risk of the importer not meeting his obligations, but there was also the danger of government expropriation and loss through political unrest. War risks were also still prevalent in many minds and had to be taken into consideration.

Apart from the obvious benefit in fostering exports by industrially developed countries, export credit finance is also meant to be a par-

ticipation and contribution in the development of the less rich countries.

The example was set by the United Nations Organization which, as soon as it came into existence, created the International Reconstruction and Development Bank (commonly called the 'World Bank') in order to finance basic infrastructure such as roads, electrical utilities, dams, irrigation, etc., and to advise and assist national development banks. The profit-making business was left to private enterprise and partly financed through export credits.

Basic elements of export credits

Although the detailed arrangements[1] are particular to each country, there are basically two elements to all export credit schemes:
- —Privileged credit terms, either direct from a government agency or guaranteed by the agency;
- —coverage of the commercial, political and other risks inherent in exporting, by either governmental or government connected agencies.

Privileged credit terms
Long term repayment period: Export credits are usually repayable over a longer period than normal bank finance would offer for a similar project in the exporting country. This was not the case to start with, for, as late as the early 1960s, export credits were limited to five years.
Privileged interest rates: Likewise, export credits are usually privileged by being available at an interest rate which is lower than that prevailing on the ordinary domestic or international money market. This would not be possible without the government's intervention either by direct subsidising or by providing funds towards a given project, such funds being combined with monies coming from the banks at normal commercial rates. A further stage has been reached in some countries which offer not only a low rate for export credits but also a fixed rate; this cannot be carried out without the participation of the corresponding government.

Coverage of risks
Most of the risks which the lender faces are covered under a credit insurance policy issued by an insurance organisation or company set up by or connected with each government. The policies are comprehensive and fairly similar in all countries, in so far as the coverage of the political risks are concerned. Furthermore, most of the credit insurance companies cover the default of payment

due to purely commercial mishaps (commercial risk). Some countries also cover, to a certain extent, the risk of termination of the contract, enabling the exporter to be reimbursed for the expenses he has incurred before his obligations have been completed.

International coordination of export credit arrangements

More and more countries, particularly the industrial ones, provided export credits in order to promote their foreign trade which had become essential to them. For a given deal, the existing technical and commercial competition between countries was to be augmented by a battle over financing terms; it soon appeared that this could not be a sound basis for the future, and various forms of international coordination and consultation have been evolved to try and halt such a trend.

Union of Bern
The credit insurance organisations of various countries formed the Union of Bern which provides for their executive officers to meet annually. During these meetings, all members exchange information and try to coordinate their individual policies, with the aim of reaching a uniformly acceptable approach to either a special type of activity — for instance, basic infrastructure, the steel industry, the oil industry — or towards individual borrowing countries. It was the Union of Bern which initially limited the repayment period of all export credit finance to five years. However, this strict limit soon appeared rather unrealistic, and the member countries unilaterally granted longer credit periods until agreement was reached within the Union of Bern on extending the credit period.

European Economic Community (Common Market)
One of the objects of the European Economic Community (EEC) was to create the best possible economic coordination between its members and, of course, exports and their financing were given particular attention. There is a special department in charge of the policy regarding export credits, and high-ranking officials of the governments of each member country meet in Brussels every two weeks to harmonise the credit terms on various projects. Fairly strict procedures have been introduced as a consequence. All members have to report to the EEC all credits approved by their governments where the repayment period exceeds five years. Objections can then be raised where too long a credit has been accepted for a given project or granted to a specific country.

Furthermore, the EEC has ruled that, when dealing with another member country, a credit longer than five years should not be accepted, long term loans being reserved for developing countries; inside the EEC, the interest terms should be those prevailing on the domestic market.

OECD (Organization for Economic Co-operation and Development)

The committee of the ministers in the OECD decided in 1963 that an advisory commission inside the organisation would deal with international coordination in the field of export credits, and the relevant guarantees. This commission of twenty-one members normally holds two and sometimes three meetings a year in Paris, with the purpose of harmonising export credit terms and thereby avoiding unfair competition. It is thus similar in purpose to the EEC organisation in Brussels. Nevertheless, the OECD's decisions do not have the same compulsory enforcement as those of the EEC; it is rather a vehicle for discussion and dissemination of information. The OECD advisory commission issues recommendations, mainly on interest rates, on the indebtedness level of the borrowing countries, and on the acceptable duration of export credits. Examples are discussed in the later section dealing with 'terms of export credits.'

Coordination for specific projects

Strict coordination between the credit insurance organisations of the EEC members has set an example to the international trading community. Now, nearly all national credit insurance organisations try to avoid erratic worldwide credit competition; before accepting that favourable credit terms be offered to a foreign country for a given project, credit insurance organisations consult each other and attempt to grant terms which are basically the same. The most impressive example of international coordination was the tentative agreement adopted by the main exporting countries in September 1974 (following the meeting of the International Monetary Fund in Washington), limiting to three years the length of the credits to be granted to developed countries or to rich oil-producing countries. However, the stagnation of world trade has prevented the effective enforcement of that agreement.

In contrast, a similar agreement which was reached by OECD in the late 1960s has been complied with (with the exception of one exporting country); ships, tankers and LNG carriers could not, as a rule, be financed through export credits for more than seven years and for more than 70 per cent of their value; 30 per cent of any payment was to be made in cash. Credits related to aircraft (including helicopters), and also nuclear power plants and steel mills, are not included in the limitations imposed by the OECD to its members regarding the length of the repayment period.

Types of export credits

The main object of export credits is the supply of finance to enable a foreign borrower — whether a private enterprise company or a governmental organisation — to carry out a given project. It also affords the exporter with recourse to a reliable source of payment. This is known as project finance. Another form of export credit is the line of credit.

Project finance

There are basically three forms of export credits available for financing specific projects: supplier credits, buyer credits, and special credits involving government participation.

Supplier credit: The first motive for export credits was the possibility for either a supplier or a contractor to grant deferred terms of payment to the buyer who wanted to carry out a project, or to order individual pieces of equipment. With this in mind a loan is granted to the national supplier, who makes the finance a part of the offer to the foreign client so that it may be more attractive. Nevertheless, the supplier has to carry the burden of that loan but then calls upon the credit insurance organisation to cover the risks incurred. However, as it is the rule that a portion of the risk should not be covered by the insurance the supplier is obliged to bear this uncovered portion; this may range between 10 and 25 per cent depending both on the policy of the exporting country and on the status of the importer. The percentage which is not covered by insurance has an impact on the prices quoted by the exporter.

Buyer credit: The idea of having a supplier in a position to quote a price on a real cash basis has led a large number of exporting countries to turn to so-called 'buyer credit'. This has become the more favoured mode of export credit financing for countries like France, the United Kingdom, Italy and the United States, whereas Germany, Japan and the Netherlands seem to have been a little more restrictive in developing this particular form.

With buyer credit the loan is no longer made to the supplier but is a credit granted directly to the foreign buyer who uses it to pay the different suppliers participating in a given project. In many buyer credit projects, the idea of real cash payments has been extended to the point where progress payments meet the effective disbursements as they are incurred during the manufacturing period.

As the exporter is being paid without delays and is assuming no risks, he is able to quote more competitive prices. In fact, when buyer credit is used, the portion that the credit insurance organisation does not cover does not lie with the supplier any longer but with the lenders; furthermore, most of the credit insurance organisations have gone one step further by reducing the uncovered portion; for example in France, COFACE covers 95 per cent of the risk in a buyer credit.

Special credits with governmental participation: The concept of export credits usually involves financing by banks and is covered by credit insurance organisations. However, in some countries, the credit is granted up to a certain amount only by the banks, while government funds are used for the other portion. In countries like Japan, the Government's Export/Import Bank finances approximately 60-70 per cent of the credit, leaving the remainder to the banks. A similar procedure exists in Italy, where Mediocredito (which uses government funds) finances the largest proportion of an export credit. In the United States, the Eximbank finances up to 50 per cent of a credit, the balance being left to the private banks; the risk for which the Eximbank may or may not cover. In another procedure in the United States, OPIC can cover the risk of the credit, leaving the whole of the financing to private banks under normal market conditions.

Another example of direct governmental participation can be found in France, where in some cases and with specific countries, the French Government has entered into a state-to-state agreement under which the French Government undertakes to use public funds for the financing of some selected projects. In such projects, the French Government finances a fraction of the deal under especially generous terms (repayment over fifteen or even twenty years with a very low interest rate), the remainder being financed under the normal terms of supplier or buyer credits. The fraction financed through government funds is normally limited to 20 per cent and only under exceptional circumstances goes beyond this percentage. These very generous terms from the French Government are linked to political negotiations, and represent the granting of special aid by the French Government to the beneficiary country.

Line of credit
Supplier and buyer credits are related to the financing of a particular project for a given foreign buyer. Experience has shown that a further step has had to be taken; the lead has been given by the United Kingdom with the line of credit. With this alternative the finance extended by the exporting country is not only meant to cover a specific project of the buyer but, more widely, all the purchases for different projects that this buyer would be making from that country over a given period.

Lines of credit clearly offer more flexibility and are likely to be developed.

Factors influencing the terms of export credits

A description is given in the appendix to this chapter of the terms normally available in a selection of countries. However. the precise terms which can be granted in a specific instance take into consideration factors such as the type of equipment involved, amounts of money involved and length of credit, and the borrowing country. Constraints may also be imposed by one or other of the organisations attempting to coordinate such matters, e.g. OECD, EEC, etc, as discussed earlier.

Type of equipment

The first aim of export credits is, as the name implies, to finance exports, and the nature and type of goods being exported influences the length of credit which might be extended. Two basic principles can be stated since they seem to apply to nearly all exporting countries:

— the credit should never be longer than the life of the equipment which is financed; a locomotive will enjoy a longer credit than an automobile;
— the more sophisticated the equipment, the longer the credit can be; an ethylene plant will be entitled to a longer credit than pipes or rails. As previously mentioned, aircraft, nuclear power plants and steel mills are allowed particularly long repayment periods of the credits. Consumer goods, even sophisticated goods, would be financed on short term credit.

The financing of services, and particularly engineering consulting services, are considered by some exporting countries as especially attractive and as a consequence these enjoy rather long term credit. Other countries would finance such engineering services only if they are associated with the provision of a fairly large amount of capital equipment.

Amounts involved/length of credit

Normally the length of credit is a dependent variable of the aggregate amount to be financed; relatively small projects ranging from, say, $1,000,000 to $5,000,000, would lead to medium terms, while larger projects for developing countries would be eligible for a credit up to ten years and sometimes, but very exceptionally, for a longer period.

Nevertheless, the type of industry concerned also influences the length of credit; an oil refinery would not qualify for more than eight years while an atomic power plant would obtain longer terms. As already mentioned, special international arrangements limit the conditions applicable to the credits covering the sale of ships, tankers and also LNG carriers. Further information on the special arrangements for government assisted financing for shipbuilding is given in chapter 6.

Borrowing country

With the second main purpose of export credits being to bring assistance to developing countries, the ruling principle should be that the less developed the borrowing country is the longer the terms of credit ought to be. This principle has been applied in one sense in that certain rich or highly industrialised countries have agreed to a self-imposed limited length of credit (evidence is given by the EEC regulations mentioned above). However, in many circumstances the influence of politics cannot be avoided, and when this is the case the rules concerning the realistic length of credit to be extended are often disregarded.

In 1976 the OECD countries decided to classify the borrowing countries in three broad categories, on the basis, essentially, of their *per capita* annual income: first category, over $3,000; second category, between $3,000 and $1,000; third category, under $1,000. For credit of two to five years (inclusive) the minimum interest rate is 7.25 per cent for all but the wealthiest category, for which it will be 7.75 per cent. For credit of more than five years the rate is to be 8 per cent for the wealthiest countries, and 7.75 per cent and 7.50 per cent respectively for those in the other two groups. Similarly, the maximum length of credit for the wealthiest countries should be five years, while for the second category the repayment period is extended to eight and a half years, and for the third category, concerning the under-developed countries, the redemption period can be ten years. Exceptions to these limitations can be granted only after consultation with the other OECD members.

Scope of financing

Export credits are in one way or another subsidised by the governments of the exporting countries and therefore have to conform to regulations which may contain certain constraints and limits and may be granted to finance certain items only.

Financing of equipment and services provided by the lending country

Its very nature prevents an exporting credit being obtained unless some goods, machinery, equipment or services are exported by the country providing the finance. These exports provide the nucleus around which export credit is arranged; in some instances the credit covers only some items included in an industrial project, in others it covers the whole of that project.

Financing of local costs

In some cases, the financing does not have to be limited to exported items and relevant services such as transportation, insurance, etc., but may also include expenses incurred in the borrowing country, if specifically approved by the export credit organisation. As this facility goes beyond the normal philosophy of export credits, the local costs eligible for financing by an export credit have to satisfy the following criteria:

i) Only erection costs and some of the civil engineering expenses can qualify.

ii) Not all costs under i) are acceptable to be financed through export credit schemes of all countries. Most lending countries establish the local cost element associated with the project to be covered by export finance, as a maximum percentage of the value of the exported equipment or services. This percentage varies from country to country and ranges from 10 to 20 per cent; the Republic of South Africa and Australia are sometimes more generous, with a total financed foreign component reaching 35 per cent.

The interest rate for the financing of such local expenses can in some countries, an example being France, be the same as the fixed interest rate of the export credit, whereas in several other countries it is the one prevailing in the money market. If the latter is adopted then the rate would be the best in that market since the loan is guaranteed by the national credit insurance organisation.

Financing of equipment or services supplied by third countries

It may have been relatively easy in the past to supply services, materials and equipment from a single exporting industrialised country, when countries able to export were few, and technology somewhat simple. Over a period of time obstacles had to be overcome; some types of equipment were not available from the main exporting country or, if they were, there was too long a delivery time or too high a price. In such cases procurement from third countries was necessary, if not essential, and consequently export credits came to include a foreign portion. This was especially the case where a key process had to be obtained from a specific country.

Export credits to cover third country supplies are not automatically obtainable and sometimes government departments need a great deal of persuasion, since they are rather careful with the use of funds they allocate for a project, especially when it may lead to a disbursement of foreign currencies. It could be said that the average foreign portion does not exceed 10 per cent, Australia and the Republic of South Africa being the most flexible in this field.

The EEC could have been a suitable entity for its members to practise a cooperative approach to financing. Actually, the Treaty of Rome provided for an allowance of 30 or even 40 per cent for procurement from other member countries. Unfortunately, whilst some Common Market countries have followed the EEC regulations, some others have deliberately decided not to be so generous and have remained within the usual 10 per cent limit.

Multinational credits for procurement from several countries

Over the years, the number of projects have increased both in number and sophistication. Consequently, the amounts involved have become greater. The stage has been reached where procuring (and therefore financing) from one single country is often no longer realistic, and multinational credits have had to be introduced. Going multinational, for large projects of course, offers several advantages: a larger variety of sources for the items, services or processes; the possibility of buying at lower costs; and a spreading of the inherent risks between the participating countries.

To be really attractive multinational procurement and multinational financing must be tackled with expertise, if they are to combine the various possibilities offered by each participating country. In each of the countries there is an optimum amount which can lead to optimum financing terms.

Multinational financing requires coordination; in the same way as on the technical side, the contractor in charge of the whole project has to make sure that no discrepancies appear due to procurement of services being spread too widely. A multinational credit has to be co-ordinated from the outset through discussions with the different governments' agencies involved, so as to make sure that the most advantageous terms can be obtained under each country's legislation. This will ensure that harmonised terms of credit are offered to the buyer.

A few companies active in export credit financing, have specialised in this multinational financing; an example is CIAVE, Paris which has developed these kind of services over the past 25 years and has created sister companies in other industrialised countries. Multinational financing seems to be the future in a world in which larger industries and new sophisticated techniques emerge every day.

Insurance costs

The governmental agencies or companies which insure export credits charge a premium which depends directly on the financial standing of the borrowing country. Most credit insurance organisations classify countries in four categories; the category which is considered to be the best, quite naturally enjoys the lowest premium. The difference between the highest and the lowest premium can be quite large. Most of the credit insurance organisations keep their categories confidential so as not to offend the national pride of any country. As the method of computing the premium is complicated, and varies from country to country, it is not easy for the borrower to know exactly in which category his country has been classified.

Some countries, notably the USA, have a more simple method of fixing the credit insurance premium. Up to 1974, Eximbank used to have a single premium of 0.50 per cent per annum, computed on the outstanding balance under the portion of the credit granted by the private banks (and not on the portion granted by Eximbank). Now the premium is fixed by Eximbank case by case and ranges between 0.75 per cent and 2.00 per cent per annum.

Brief survey of characteristics pertaining to various countries

The appendix to this chapter tabulates the various characteristics of export credits available from a number of countries. To supplement this a brief summary of the general structure of credit insurance and of the financing systems prevailing in the main exporting countries is given below. Details reflect the situation as of January 1977. Of course, all the figures mentioned and the procedures described in the following pages have to be considered as a general view of the prevailing arrangements. As such they are subject to change at any time (as has been the case in the past) with the trend towards international harmonisation.

United Kingdom
Export credits from the UK are guaranteed by the Export Credits Guarantee Department (ECGD), a unit of the UK Department of Trade. This was created in 1919 and supported by the UK Treasury with money being voted annually by Parliament and surpluses being returned to central government funds at the end of each year. The finance for the credits covered by ECGD is provided by the British clearing banks. Nevertheless, *many of the buyer credits for large investment projects are arranged by the merchant banks.* Both supplier and buyer credits are available. The interest rate for UK credits is now fixed and officially ranges between 7.25 and 9 per cent plus current banking charges. However, 8 per cent is in fact the effective minimum rate for credits over five years. In order that this rate be granted for export credits, the British Government (through

the Bank of England) reimburses the clearing banks for the difference between certain observed domestic rates and the rate effectively quoted for export. The United Kingdom has also set up a mechanism of progress payments during the manufacturing or the construction period, enabling pre-shipment payments. On buyer credits, ECGD covers 100 per cent of all political and commercial risks both for so-called private or public buyers. For supplier credits after shipment, the coverage is for 95 per cent of political risks and only 90 per cent of commercial risks.

In February 1975 the United Kingdom introduced an insurance scheme against escalation of exporters' costs for large capital goods contracts with manufacturing periods of two years or more. Such a guarantee does not cover the first 10 per cent per annum of increases in costs. The procedure was improved in November 1975 and considerable pressure is being brought by industry to make further improvements.

It is possible to have credits in currencies other than sterling under ECGD coverage. Recently ECGD was authorised to improve the cover afforded for such contracts which are expressed in certain major foreign currencies, and are supported by use of the forward market or foreign borrowing. This was intended to bring it more nearly into line with the cover provided under sterling contracts. Credits in dollars under ECGD coverage are most likely in future to represent a large part of UK credits under ECGD procedures.

Federal Republic of Germany
German export credits are covered by Hermes Versicherungs AG (generally referred to as 'Hermes') which is a private insurance company operating on behalf of the German Government. Political, foreign exchange and commercial risks are covered. These risks cannot be covered separately. Decisions concerning the terms to be granted are made by a special committee in which representatives of various Ministries of the West German Government sit together with representatives of the German banks and industries.

There are two organisations involved in this but the mechanisms of export credits are fairly complicated:
AKA (Ausfuhr Kredit Anstalt)
The goal is to participate in export financing on medium and long terms. The prevailing financing is made through AKA, a private company created in 1952 with the participation of 52 German banks. Three lines of available funds exist with corresponding ceilings of allocations for any project. These are referred to as *line A, line B and line C* as detailed below:

> *Line A:* (Maximum available finance: 6.5 billion deutsche marks.)
> The German exporter's bank will grant directly 40 per cent of

the credit (the balance of 60 per cent coming through the banking pool of AKA), the total credit being limited to the percentage of the contract covered by the Hermes guarantee. Such credits are made available to the exporter progressively during the manufacturing period. Export credits financed through *line A* are not too frequent. They apply mainly to medium and long term credits.

Line B: (Maximum available finance: 3 billion deutsche marks.) This line has a right of discount with the central Deutsche Bundesbank. Credits under this line are 70 per cent of the total amount of credit. The down payments are not included under this line and the German exporter can currently finance 50 to 55 per cent of the contract value. The funds are made available to the exporter through his bank which draws on *line B* credit from AKA.

Currently a combination of financing through *line A* and *line B* operates most frequently.

Line C: This line was set up in 1969 by the banks participating in AKA. 4.5 billion deutsche marks have been transferred from amount available of *line A* to make available a facility for so-called buyer credits i.e. *lines A & B* are supplier credits. Actually this wording does not correspond to the definition of buyer credits in other countries. The mechanism of *line C* is that AKA acquires the promissory notes held by the banks corresponding to the export for which all the obligations of the suppliers have been fulfilled and the technical guarantees have been satisfied. Nevertheless, the supplier is left with 20 per cent of the risk of non-payment of all outstanding instalments.

The interest rates applicable to the above mentioned AKA credits fluctuate. For *line A,* it is variable depending on the money market conditions (7 per cent since October 1976). For *line B,* it is always 1.50 per cent over the discount rate of the Deutsche Bundesbank, and for *line C* a 7.25 per cent per annum rate was enforced in 1976 but variable depending upon the money market conditions. Additional variable financing fees charged in normal banking practice have to be added to these interest rates.

It is important to observe that all the above-mentioned privileged interest rates apply only to the portion financed through the above-mentioned agencies. They leave with the exporter a substantial part of the credit—consequently the exporter has to consider the impact on his pricing of the cost of funds which are not provided by the above-mentioned specialised export credit agencies. As a consequence, the interest rate offered to the foreign buyer is substantially higher than the existing rates for the funds provided through AKA and KFW.

KFW (Kredit Anstalt für Wiederaufbau)

This financing organisation was set up in 1948 by the German Government mainly to re-establish German basic industries. In addition credits are given for long term export financing to developing countries. More recently the interest rate of KFW was increased for normal export credits financed through KFW. Since the second half of 1976 it has been 8 per cent per annum for supplier credits and buyer credits. The credits are granted to German suppliers and foreign buyers only via other banks. All KFW credits have to be covered by mortgages and/or the Hermes insurance and are restricted to important projects. KFW has granted credits on long term bases sometimes reaching twenty years or more when the projects were contracted by foreign governments and represented an important development of the foreign country's economy.

KFW credits are available only after the start up of the export installations, so that interim financing has to be arranged. This financing is often covered by the commitment of KFW to finance the project later on.

France

Credit insurance was set up prior to the Second World War. The first French credit insurance company was 'Société Française d'Assurance pour Favoriser le Crédit'. In 1946 the Compagnie Française d'Assurance pour le Commerce Extérieur (COFACE) was created under tight governmental control with government backing to cover all medium and long term risks which were incurred. COFACE covers mainly export credits. In addition, special guarantees can be requested to cover cost escalation, exchange rate losses and investments. The latter three special insurances have restrictive regulations and are not fully comprehensive.

For short term credits up to two years, a 'global' insurance policy is granted to the exporter, covering compulsorily all qualifying exports of said company. Coverage of 80 per cent is available under this type of policy for both political and commercial risks extended to private and public buyers. The interest rate for such credits is the prevailing rate on the domestic market.

For credits over two years, both supplier and buyer credits are available. Normally 85 per cent of the risk is covered for supplier credits to so-called private foreign buyers, and 90 per cent for so-called public buyers enjoying a foreign governmental or a specialised governmental bank or agency guarantee. The risk covered for buyer credits is 95 per cent.

The regulations set up by the agreements reached among the OECD countries in June 1976 and discussed previously in the section dealing with 'Terms of Export Credits., raised the French interest rates to a higher percentage than those charged up to that time. France is the only country in which the minimum rates agreed upon by the OECD correspond to the actual rates, while in other countries, in practice, the rates are higher (see Appendix at the end of this chapter). The interest rates for credits of two to five years are now 7.25 per cent, for all but the wealthiest countries; these are charged at the 7.75 per cent rate. For credits over five years, the rates are 7.5 per cent per annum and reach a maximum of 8 per cent for the wealthiest countries.

The French export credit interest rate is fixed for the life of the contracts during the whole reimbursement period.

Export credits are granted by the banks for credits up to seven years; the promissory notes being endorsed by Banque Française du Commerce Extérieur. Re-discount facilities at a lower rate with the Banque de France are enforced by regulations for part of the instalments maturing during the first seven years. For the instalments maturing after seven years, the credits are made available by Banque Française du Commerce Extérieur, which can call on funds provided by the government.

Pre-shipment financing is available and a progress payment mechanism has been set up during recent years at the same, low fixed interest rate as for the main credit. As progress payments are strictly controlled by the Banque de France, in order to avoid too early payments to the exporter, the latter can have the benefit of a so-called 'pre-financing' credit, available at a special interest rate which is higher than the normal export credit. The said rate is officially fixed from time to time.

It is worth noting that long term financing is based on the idea of favouring the developing countries. Consequently, such terms are not eligible for developed countries, including the EEC countries. An exception to this restrictive regulation exists for North Sea offshore equipment, for which the prevailing terms are: 20 per cent down-payments, reimbursement in 10 semi-annual instalments, interest rate 7.75 per cent per annum.

United States of America

There are four organisations concerned with various aspects of export credits in the United States, but by far the majority of credits are set up through Eximbank.

Eximbank (Export Import Bank of United States)

Prior to 1974, Eximbank's credits were generally provided on the following basis:

   —10 per cent down-payments;
   —45 per cent financed directly by Eximbank; such credits

corresponding to the instalments maturing during the later redemption period at a fixed interest rate of 6 per cent;
— the balance of 45 per cent financed by American banks (and sometimes even by non-American resident banks) at the normal fluctuating market rate of interest, the instalments of this latter credit maturing during the first half of the redemption period and guaranteed by Eximbank for a premium of 0.50 per cent per annum.

The rapid increase of Eximbank's commitments between 1969 and 1974 was criticised by the American Congress, and more restrictive terms have since been imposed. The prevailing terms require a down-payment of 15 per cent (compared to 15 or 20 per cent normally requested as down-payments by other exporting countries), and Eximbank's financing has been limited to only 30 per cent of the credit. It should be noted that recently and in certain situations the former 45 per cent financing might again be accepted by Eximbank. The balance of financing must be obtained from normal banking sources, and the coverage of the risk for the whole amount of the credit is not always the practice. A certain amount of risk is left with the banks. The mechanism of reimbursing the banking credits prior to the Eximbank credits is normal practice. The interest rate in the new Eximbank scheme is higher and in January 1977 varied between 8.25 per cent and 9.50 per cent per annum depending upon the length of the credit.

The Eximbank guarantee fee has also been increased and varies between 0.75 per cent per annum and 2 per cent per annum depending upon the view taken as to the solvency of the borrower. It is computed only on the portion of the loan extended by the banks (excluding the Eximbank).

PEFCO (Private Export Funding Corporation)
PEFCO was set up in 1967 by the United States Bankers' Association and is today capitalised by fifty-five major American banks and several industrial corporations. PEFCO issues its own debt securities to raise funds and uses this money to purchase foreign debt obligations related to American exports. The promissory notes purchased by PEFCO must be guaranteed by Eximbank. PEFCO credits have the benefit of a fixed rate which in the second half of 1975 was approximately 9.25 per cent.

FCIA (Foreign Credit Insurance Associations)
This organisation grants insurance policies to the exporters with the support of Eximbank. Under the coverage of such credit insurance, American banks can finance an American exporter at the fluctuating market rate of interest. FCIA coverage provides protection against loss both for political and commercial risks.

OPIC (Overseas Private Investment Corporation)
OPIC was formed in 1969 by an amendment to the US Foreign

Aid Act. OPIC seeks to stimulate investments in under-developed countries by providing insurance on equity capital, and makes loans and provides loan guarantees in order to overcome financial and political risks.

The aim of OPIC is to help the development of the economy of the host country while eventually aiding exports from the United States.

Japan

Export credits are insured in Japan by a special division of the Ministry of International Trade and Industry (MITI).

The mechanism of Japanese financing is based on credits granted to the Export Import Bank of Japan. This institution generally finances 70 per cent, and on rare occasions 80 per cent, of the credit granted (45 per cent for ships). The financing provided by the Export Import Bank of Japan bears a fixed interest rate. The latter was for a long time in the region of 6 per cent. Recently the average rate has ranged between 7 and 8 per cent. The Japanese banks finance the balance of the credit and the blended rate of Japanese export credits for medium to long term ranged between 7.50 and 8.50 per cent during 1975.

It is noteworthy that MITI appears to be quite flexible on financing foreign components included in orders for Japanese equipment.

References

[1] The description of export credit schemes and of the related terms given in this chapter are those prevailing in early 1977. All matters concerning export credits are altered frequently either following changes in regulations made by the government agencies or due to the condition of the money market especially relating to the level of interest rates.

APPENDIX

Table 18: Summary of export credit financing schemes of various countries

Country	Down payments during manufacturing period	Percentage covered by credit insurance[2]	Length of credit redemption period[10]
United Kingdom	15 to 20%[1]	B.C.* 100% S.C.** 90 to 95%	Up to 10 years
France	15 to 20%[1]	B.C.* 95% S.C.** 85 to 90%	Up to 10 years
Germany	15 to 20%[1]	80 to 85%	Up to 10 years
Italy[11]	15 to 20%[1]	Exceptionally 95%	Up to 10 years
Belgium	15 to 20%[1]	B.C.* 90 to 95% S.C.** 80 to 85%	Up to 10 years
Holland	15 to 20%	B.C. 90 or 95% S.C.** 75 to 90%	Up to 10 years
Spain	B.C.* 15 to 20% S.C.** 20%	B.C.* 90 to 95% S.C.** 80 to 85% commercial risk 90 to 95% public buyer	5 years Exceptionally up to 10 years
Japan	15%	90%	Up to 10 years
United States	10%	30 to 100%[8]	Up to 10 years
South Africa	15%	B.C.* 100% S.C.** 90%	Up to 10 years
Australia	Minimum 10%	B.C.* 100% S.C.** 90 to 100%	5 to 10 years
Switzerland	Variable 10 to 30%	70 to 95% Case by case	Up to 10 years

*B.C.: Buyer Credit.
**S.C.: Supplier Credit.
[1] 30 per cent for ships and rich country borrowers.
[2] In certain cases these percentages might be reduced.
[3] Banking and Credit Insurance charges not included.
[4] Fixed interest rate during life of credit.
[5] Do not exceed amount of down payments.
[6] Export credits over five years are not currently granted. Normal rates are 9 to 9.50% but various forms of subsidising allow lower rates.
[7] Credits for local costs carry an interest rate between 6.25 and 7.50% per annum, and might have to be reimbursed over a shorter period than a long term export credits.

Interest rate[3]		Local costs guaranteed up to	Financing of exporter during manufacturing period through export credits	Guarantee of cost escalation
Medium term %	Long term %			
7.25-7.75[4]	7.50-9 Normally 8[4]	15%[5]	Yes	Possible
7.25-7.75[4]	7.50-8[4]	15%[5]	Yes	Possible
7.25-8.5	7.5-8.5	15%[5]	No	No
Minimum 8.50	7.75-9	Exceptionally[5]	Exceptionally	
7.25-9.30	8.25-9.80	15%[5]	No	No
7.25-9[4]	7.75-9.50[6]	15%[5]	No	No
B.C.* 8.10 S.C.** 7.50[4]	8.10 7.50	20%[5]	Yes	Possible
7.5-8.5	7.5-8.5	15%[5]	No	No
9-10.5	8.75-10.50	No	No	No
7	7.5[4]	35%[9]	Yes	No
Fluctuating		35%[9]	No	No
5-6.5	$6\frac{1}{8}$-$6\frac{7}{8}$	25%[7]	No	No

[8]100 per cent coverage is now exceptional.

[9]Including possible financing of procurement from various other countries.

[10]The repayment of the credits normally starts six months after a contractual date which is determined case by case by the government's credit insurance organisations. This date is at the earliest the time of delivery of individual equipment and at the latest the date of the whole plant commissioning. (See section in text entitled 'Factors influencing the terms of Export Credit'.) As mentioned in this chapter credits for ships shall not exceed seven years while aircraft, nuclear power plants and steel mills can be reimbursed over more than ten years.

[11]Quotations and reimbursements in US dollars are normal practice due to prevailing conditions since 1976.

Further Reading

Documentation on export credits is fragmentary and depends on governmental regulations and banking practice, some of which change quite frequently. A list of addresses for the principal countries from which documents are issued periodically is given below. With the exception of the final item, the remainder relate to the scheme of the particular country. The list is not exhaustive. Other countries also publish information.

United Kingdom:
Export Credits Guarantee Department 'ECGD Services',
Marlon House, Mark Lane,
London EC3R 7HS

France:
Compagnie Française d'Assurance pour le Commerce Extérieur
(COFACE)
'Assexport' (Restricted circulation)
32, rue Marbeuf,
75008 Paris

Germany:
AKA (Ausfuhr Kredit Anstalt Gesellschaft mbH)
Frankfurt

Spain:
Banco Exterior de Espana,
Carrera de San Jeronimo 36,
Madrid 14

Banco de Vizcaya,
Alcala 45,
Madrid

USA:
Export-Import Bank of the United States,
811 Vermont Avenue N.W.,
Washington D.C. 20571

Australia:
Export Finance and Insurance Corp. (EFIC)
(An Australian Government Statutory Authority)
2, Castlereagh Street,
Sydney

The National Bank of Australasia Ltd.

International:
Documentation published by
The Banker Research Unit
(A division of The Financial Times Ltd)
'The B.R.U. Export Finance Service' (Quarterly Reports)
10, Cannon Street, Bracken House,
London EC4P 4BY.

12

Stock Markets

Petroleum companies of all sizes increasingly recognise that the raising of finance is now very much a matter to be undertaken on an international basis and that the stock markets of the world, together with their specialist capital-raising associates, are substantial sources of funds.

To be able to use these markets effectively it is essential to recognise that whilst all markets are similar in nature there are important differences of detail, particularly in terms of the financial capacity of the market, requirements for listing, method of operation, means of regulation of the market and possibly specialist interests.

Market organisation

The organisation of stock markets can be conveniently considered in two parts:
- the method of operation of the stock exchange
- the regulation of market activities

At the outset it is important to stress two important aspects relating to all stock markets. Firstly that stock exchanges and their members are not in themselves a source of funds. Rather do they act as a catalyst in converting liquid resources into working capital and the means of acquiring real assets. Secondly that none of the stock exchanges, nor any of the regulatory bodies, purports to pass any judgment on the merits or otherwise of the shares or companies which are the subject of dealings or offerings. They seek merely to ensure, as far as possible, that adequate information is disclosed (and equally that nothing relevant is omitted), that any suspect promoters or known offenders are barred and generally that investors are protected against everything — except their own failings.

The development of a centre as a major capital market turns on a number of factors. Generally speaking a high level of domestic wealth creation is necessary to provide the funds, complemented by a firm political and legal framework and financial institutions of undoubted integrity. In more recent times the domestic level of available funds has tended, in some areas, to be augmented by the import of foreign funds, sometimes for re-export, and this form of financial entrepot trade ideally requires a stable level of exchange rates if shocks are to be avoided on interest and repayment schedules.

Stock exchanges

Exchanges have existed since man first moved away from self-sufficiency in his requirements. At first they were simply general markets where goods were bartered or exchanged. In time they developed from barter to cash transactions and tended to specialise in certain commodities such as wool, cotton, corn and so on, or even in intangible items such as insurance or ship charters. More recently markets have developed which do not require the physical presence of either party to the transaction. The foreign exchange market, for example, has no physical existence in the sense of being situated in a building; it is made up of a multiplicity of dealers, each in his office and conducting the business by telephone.

Stock exchanges are no different from any other form of market in that they exist to provide dealing facilities in a specific range of items, namely the shares, stocks, debentures and securities of all kinds, not only of incorporated entities but also of governments, public corporations and other institutions. Stock exchanges have often been the subject of much adverse comment but this is attribut-able very largely to ignorance of their place in an advanced financial society. Their role is solely to provide a means of bringing buyers and sellers together. The buyer will be seeking to utilise funds at his disposal to invest in the security of his choice, whilst the seller will be seeking to convert an existing holding into cash. The stock exchange provides the facilities for the two parties each to achieve his aim.

The securities which form the subject of the majority of stock exchange transactions are, of course, already in issue and held by an individual or institution. This fact suggests at first sight that stock exchanges do not play a role in the primary market of raising new and additional money, but this is a purely superficial view. Except in special circumstances, such as the acquisition of trade investments by a company, the investor is concerned to know that he will be able to realise his investment when required, not only in the event of needing cash, but also to facilitate an investment policy which will require that holdings are changed from time to time. Stock exchanges not only provide the facilities whereby this requirement is met but also, through the knowledge and experience of the members who make up the exchange, the mechanism whereby offers of new securities are underwritten or placed with buyers.

The role of stock exchanges in providing a primary source of funds is thus very real and depends upon confidence and availability of funds for investment.

Market regulation

Virtually all countries now have comprehensive legislation governing the establishment and running of corporate entities, but the means whereby stock markets are regulated, and shareholders protected, vary from country to country.

The major stock exchanges of the world ensure that varying degrees of disclosure, but nevertheless adequate information is made available for the benefit of intending investors, and in this respect their self-imposed regulations often go beyond what is required by corporate law. Stock exchanges also have comprehensive rules in regard to requirements for listing, dealing procedures and so on, and there is often some form of compensation fund to make payments in respect of losses arising from default by exchange members.

These developments by the stock exchanges themselves may be partly in recognition of the inevitability of state supervision if mal-practices are rendered possible due to lack of adequate control, but the motivation is unimportant in relation to the outcome, which is that ethical standards and the scope of information disclosure demanded by most stock exchanges are higher than ever before. In spite of this many countries have instituted some form of government supervision as discussed in the following sections dealing with individual markets.

The London Stock Market

The London Stock Exchange has been in existence for almost two hundred years and currently has some 4,500 members who are registered either as brokers or jobbers. The brokers act as agents for the public and deal with the jobbers who act as principals in making a market in a range of shares usually classified by industry. The jobber will normally quote two prices, the lower being the price at which he will buy shares and the higher the price at which he will sell shares. The difference, the 'turn', represents his profit. It is rarely as simple as this, however, and the turn may increase or disappear, depending on the course of markets. The jobber therefore needs a very agile mind to keep abreast of developments. The broker basically has the problem of dealing at the most satisfactory price for his client, from whom he receives a commission as his remuneration. There are many subtle aspects of the market relationship but it is generally agreed that the broker-jobber system provides a very sophisticated form of marketplace.

The majority of dealings are concerned with shares which have been officially admitted to listing and achievement of this status requires compliance with far-reaching provisions in respect oı disclosure of information, not only at the time of listing, but also in respect of undertakings for future disclosures. It is also possible to deal in shares which are listed on recognised overseas exchanges and, under certain conditions, in the shares of companies which are not officially listed. The significance of this latter provision is discussed later.

In Britain, whilst the Department of Trade has far-reaching powers in regard to companies there is no official regulatory body. A major safeguard is undoubtedly the high ethical standards of the members of the professional bodies of accountants, brokers and solicitors, and of the various city institutions. In addition, the Panel on Take-overs and Mergers, set up by the various financial organisations in the City, monitors the activities indicated by its name. This rather curious British compromise works very well, perhaps largely because the professional standing, and indeed livelihood, of those concerned depends on business being conducted in a manner which is above criticism.

The New York Stock Market

There are three separate markets in New York. The New York Stock Exchange (the 'Big Board') is by far the largest, and lists the majority of the leading American companies. The American Stock Exchange lists mainly the smaller American companies, whilst the Over-the-Counter market lists a curious mixture of small companies, bank and insurance shares and US Government bonds. The Over-the-Counter market has virtually no listing requirements and thus provides a market for those companies which are unable or unwilling to meet the full disclosure requirements of the two principal exchanges. In addition to the markets in New York, there are trading floors in a number of other cities, and taken as a whole, the American stock exchanges have a turnover exceeding that of all the other major exchanges in the world put together.

The dealing procedures on the two principal exchanges are broadly similar to those operating in London with a market being made in a group of shares by a specialist. The analogy between the specialist and the jobber must not be taken too far, however, because the specialist in New York lacks the absolute freedom as to price variation enjoyed by the London jobber. Furthermore the very sophisticated communications system in use in New York ensures that full details of every transaction are disclosed immediately, whereas in London the jobber's book is in many senses a closed one.

The prime and original regulatory body is the United States Securities and Exchange Commission (the SEC) which came into being after the stock market crises in 1929 and the early 1930s, and which has a very powerful position indeed in every aspect of the securities trading business. It could be mentioned in passing that such close regulation may have adverse effects, in that the time and expense involved in meeting SEC requirements often inhibits overseas companies from clearing details of their share offerings with the SEC, thus precluding American investors from participating in offerings of securities by foreign companies.

Continental European Markets

By comparison with New York and London the stock exchanges of continental Europe are much less active and do not normally provide continuous dealing facilities throughout the working day.

In France the Paris Bourse handles all but a small part of the country's overall trading in shares. The brokers are public officials appointed by the state, and business is conducted in short daily sessions by fixing a price which will enable the maximum number of sales and purchases to be matched. It is not permitted that the price should vary from that ruling previously by other than a small percentage, and dealings which take place outside the official trading period must be concluded at the last official price. There is thus little or no continuous adjustment of price to facilitate dealings and the sale or accumulation of a substantial number of shares may take some time.

The Zurich Stock Exchange is the largest in Switzerland but there are other important exchanges in Basle and Geneva. The membership of the exchanges consists very largely of the local banks, and dealings are matched as between buyer and seller.

There are eight regional stock exchanges in West Germany, of which that at Frankfurt is the most important. Various financial institutions act as brokers, but the market is maintained by specialists who are appointed by the German Finance Ministry.

Holland's principal stock exchange is that in Amsterdam. The membership consists of both private individuals and financial institutions, and dealings are on the basis of matching sale and purchase orders. There are, of course, a number of very substantial Dutch companies such as Philips and Royal Dutch Petroleum; the shares of which are the subject of international arbitrage, and dealings therein are accordingly on a rather different basis.

Stock market regulatory arrangements in Europe vary; France has its Commission des Opérations en Bourse, which is broadly similar to the SEC in the scope of its operations, whilst Germany and Switzerland have no central regulatory agency at all.

Rest of the world
Apart from the countries already discussed, the others of significance are Canada and Japan.

The Toronto Stock Exchange accounts for some 70 per cent of the value of the total dealing volume in Canada, whilst Montreal handles a further 20 per cent. The balance is accounted for by various other exchanges of which mention can be made of Vancouver, which specialises in the smaller mineral companies, and Calgary, which has a special interest in oil stocks. The Toronto Stock Exchange has no specialists or jobbers but operates by a system of trading posts, each covering a group of shares. An official of the Exchange at each trading post accepts specific bids for, and offerings of, shares and, in so far as they match, deals are concluded. The trading posts do not make a market in the sense that they perform any function other than recording bids and offers. To a degree this form of market must be less flexible than the types mentioned previously, but in practice the system functions satisfactorily and brokers are able to utilise their experience of the procedures to carry out their clients' orders. The immediate disclosure of the details of each transaction follows the American pattern very closely.

In terms of market regulation the position in Canada tends to be even more complicated than in America, in that regulation is a Provincial matter (rather than a Federal jurisdiction) and hence each Province has its own Securities Commission.

Of the eight stock exchanges in Japan that in Tokyo is by far the largest. The market operates by a somewhat complicated system of matching buying and selling orders, and is characterised by the domination of a small number of brokerage houses. The Ministry of Finance plays an important part in the regulation of Japanese markets by enforcement of the Securities Exchange Act of 1948.

International listing of securities
As would be expected, the initial listing of securities usually takes place in the country of domicile, but there is an increasing tendency for companies to seek listings on foreign stock exchanges. The motivation for such moves has varied; the earlier foreign listings arose from companies which conducted substantial business overseas, e.g. British tin mining companies seeking a listing in Malaya, or where historical reasons of financing gave rise to numerous foreign shareholders, for instance the Royal Dutch Petroleum Company having many British shareholders. More recent listings have tended to arise for rather different reasons, mainly the desire to promote an international image, or to become known in an overseas centre which is a potential source for future capital and loan raising activities.

Stock exchanges, quite naturally, are always willing to add reputable companies to their list, but there are many problems which arise; mainly in the field of meeting local requirements in accounting methods and the degree of disclosure, but also in the realms of exchange control and treatment of taxation on profits and dividends (see chapter 9). The rules of regulatory bodies must also be considered, and the feasibility of an overseas listing will often turn on the degree to which the appropriate regulatory body is willing to be flexible in its requirements.

Some idea of the extent of listing of overseas companies on some of the principal stock exchanges may be obtained from the figures in table 19, which relate to the position at the end of 1975.

Table 19: Listed equity issues

	Total number of companies listed	Percentage foreign-registered companies
Japan	900	1.6%
USA	2,900	3.7%
Britain	3,800	10.5%
Netherlands	675	51.3%

It is important not to read too much into these figures, especially those for the Netherlands, which have a long history as an international trading centre, but there may be some significance in the figures for the United States and Japan, where the powers of the regulatory bodies are very far-reaching.

Comparison of world stock markets

The New York Stock Exchange is still by far the largest and most influential of world stock markets. While the established stock exchanges in the United Kingdom and North America have merely maintained their size, after adjustments for cyclical market trends, the exchanges in continental Europe and Japan have continued to expand, as reflected in table 20.

These growth markets remain individually relatively small, by comparison with the New York Stock Exchange, but in combination now represent a significant proportion of the total capitalisation of world stock markets. The trend towards rapid development in these exchanges appears to be more extreme in respect of dealing volume, than in terms of market capitalisation, as illustrated in table 21.

Table 20: Average capitalisation of exchanges
Expressed in billions US dollars

Stock Exchange		1975	1974	1972	1970	1966
UK	(London)	74.5	60.4	130.1	75.2	65.8
France	(Paris)	33.8	28.8	33.7	22.2	18.9
Germany	(All Exchanges)	50.0	43.6	44.3	31.4	18.3
Netherlands	(Amsterdam)	14.5	12.9	14.3	12.4	8.8
Switzerland	(Zurich)	18.0	14.0	17.0	10.0	n.a.
USA	(New York)	658.6	604.8	810.6	569.6	510.3
Canada	(All Exchanges)	51.4	63.3	57.5	51.1	43.4
Japan	(1st Section)	134.0	129.6	108.2	44.3	23.1

Source: Royal Trust Company of Canada

Table 21: Value and Growth of Equity Turnover

	(US$ billion)					(%)		
	1975	1974	1972	1970	1966	1975/74	1975/71	1975/66
UK	19.6	14.8	25.1	10.6	5.0	32	20	291
France	7.3	5.3	8.3	4.0	2.5	38	74	201
Germany	11.2	5.1	7.0	3.3	1.1	117	149	881
Netherlands	2.5	1.9	3.0	1.5	0.3	36	74	908
Switzerland	20.0	15.0	16.0	7.0	n.a.	33	66	n.a.
USA	133.7	99.2	159.7	103.1	98.5	35	−9	36
Canada	5.4	6.2	8.4	4.7	3.6	−13	−13	51
Japan	51.2	41.6	68.2	24.3	14.0	23	32	266

Source: Royal Trust Company of Canada

The London Market
One of the most active capital markets is in London, and details of the amount of new money raised by the issue of securities with a listing on The Stock Exchange — including the offer of new shares and debt for sale and the issue of shares by way of 'rights' to existing shareholders, but excluding issues by the British Government and public bodies — are set out in table 22.

Table 22: London Market
Expressed in millions pounds sterling

	1976	1975	1974	1973	1972	1971
Debt	93	213	43	43	295	341
Preferred & ordinary	1068	1366	119	168	661	323
Total	1161	1579	162	211	956	664
Issues by oil and mineral companies	est. 95	78	1	11	1	141

Source: Midland Bank Limited

It will be seen that, so far as listed securities are concerned, the London capital market is very much equity based; a situation which arises from the considerable equity involvement of insurance companies and other financial institutions, coupled with fiscal considerations.

The figures for 1971 include the massive operation by British Petroleum which raised £123,000,000 from the offer of new ordinary shares to its own shareholders by way of 'rights', whilst those for 1976 include £15,000,000 of 7 per cent Convertible Preferred shares for Ultramar and a substantial £76,000,000 for the joint offering by London & Scottish Marine Oil Co. Ltd. and Scottish Canadian Oil & Transportation Co. Ltd. This latter operation raised funds for the companies' commitments in financing the development of the Ninian field in the North Sea, and took the form of £75,000,000 of 14 per cent Unsecured Loan Stock 1981/83, accompanied by the unique units of Oil Production Stock, which carry an entitlement to receive half-yearly payments directly related to the market value of production from the Ninian field.

The New York Market
As befits the largest capitalist economy in the world the New York capital market is enormous by any standards. It would be unwise to attempt too direct a comparison with other centres but figures for the public corporate sector in recent years are as shown in table 23.

Table 23: New York Market

Expressed in billions US dollars

	1976 est.	1975	1974	1973	1972	1971
Net issuance of corporate domestic and foreign bonds	36.9	39.0	29.7	14.5	19.9	25.6
Corporate stock issues — common & preferred	13.1	10.3	4.3	9.1	13.0	13.5
Total	50.0	49.3	34.0	23.6	32.9	39.1

The most obvious point which stands out from these figures is that, as compared with London, New York is much more concerned with bonds than with equity. The explanation again lies with the traditional pattern of investments adopted by the financial institutions, and the degree of concern with bonds is emphasised by the very sophisticated system for grading this type of security which has developed in the United States.

Raising New Capital

Bank finance does not concern us here and hence the fundamental determination required from a company in need of funds is between debt and equity financing. The two main considerations are the existing debt/equity ratio pertaining to the company, and the nature of the project to be financed. To take elementary examples: the financing of a purely exploratory drilling programme by a small company, is obviously most suited for financing on an equity basis; the development of a proven field, or the construction of a pipeline or refinery with a determinable income flow, may be more appropriately financed by debt with a coupon and repayment programme geared to forecast income.

Having decided on the form of finance, i.e. debt or equity, the company must seek advice on the most suitable source of money. One approach is to consider raising the money from existing shareholders, and in some countries, for example, Britain, virtually all new equity offerings are made initially to shareholders by way of 'rights', i.e. an entitlement to take up new shares in proportion to an existing holding at a price substantially below the ruling market price. Generally speaking, a market will develop in the 'rights' and existing shareholders who are unable or unwilling to subscribe will sell their 'rights' in the market. In other areas, the United States for example, it is common practice to issue treasury shares, i.e. shares authorised but unissued, to outside parties at a price only marginally below the current market price.

In the first instance, the success of a 'rights' issue depends upon the existing shareholders taking up their quota of the issue in proportion to their existing shareholding. If for any reason this is not possible, it is then necessary to establish a price for the shares, and unless they are listed on one or more of the exchanges, or a market exists in the shares, the determination of a price acceptable to all parties can be very troublesome.

London market requirements

Access to the very substantial market which exists in London is relatively easy for companies which are already listed on The Stock Exchange and requires merely the up-dating of information already supplied, and an explanation of the purposes to which the new money is to be put. The requirement for companies new to The Stock Exchange is that a full prospectus be prepared and made available. This has not only to meet the requirements of the Companies Acts but also the fuller and more detailed conditions of The Stock Exchange.

It is these latter conditions which are of major concern to companies engaged in the petroleum industry because The London Stock Exchange has taken a very firm line on the listing of oil exploration companies. (The section governing the admission of securities of companies which are engaged in mineral exploration is set out as an appendix to this chapter.) Basically their requirements state that for a company to have a quotation in London it must have sufficient income from existing operations to finance its known exploration commitments or be able to produce expert opinion on its ownership of proven reserves, and estimates of the cost and time involved in bringing the company into a revenue-earning position. This effectively means, of course, that the initial stages of exploration and development cannot be financed through The Stock Exchange unless it is undertaken by a company with other substantial business. The Stock Exchange's stand on this matter has been an endeavour to protect the unsophisticated investor. The critics point to the long history of The Stock Exchange in financing mineral developments all over the world and also to the interesting paradox in the present situation in that if a company is quoted on another stock exchange it can be traded in London under rule 163.1(e). An example of this is Southern Pacific Petroleum which is an Australian oil exploration company with interests in the North Sea, quoted on the Australian Stock Exchanges and freely traded on the floor of The London Stock Exchange, although it does not have an official quotation.

The opposite school of thought bases its arguments on the current climate of regulation and control, actual or envisaged, and feels that The Stock Exchange has a duty not to facilitate investment in companies with few tangible assets but just hopes and ambitions in respect of unproven acreage.

In turning to listed companies, Oil Exploration Ltd, which has an interest in the Hewitt field and a number of other licences, is probably the best example of what The Stock Exchange is trying to achieve. The basic value of the company can be determined by its income arising from the sale of gas, which in turn more than covers its exploration commitments. Another example is Premier Consolidated, which acquired production in the US, producing income just sufficient to finance its exploration commitments and in turn satisfying The Stock Exchange requirements.

To overcome the problem of listing for oil exploration companies a number of member-firms of stockbrokers (as distinct from The Stock Exchange itself) have been active for a number of years in placing the shares of companies engaged, directly or indirectly, in oil exploration. The shares of these companies are not listed, but are nevertheless the subject of dealings under the provisions of the rules which permit dealings in unlisted securities, and there has been considerable activity in these shares from time to time. Sea Search is an example of a company put together prior to the 1972 round so as to participate in a consortia application. The company raised £120,000 with the right to call a further £1,000,000 if and when it required funds to fulfil its share of the exploration programme.

On the other hand, Siebens Oil & Gas is an example of a company which came to the market after licences were awarded to raise £15,000,000 mainly to finance its exploration programme.

Another issue of interest was that of Cambridge Petroleum Royalties which sought £2,000,000 to purchase royalty interests from operating companies in the UK and Irish sectors.

These three companies are registered in the UK and their private issues were primarily subscribed by UK institutions, with no quotation for the securities.

To meet the situation The Stock Exchange has recently amended its rule 163 which governs dealings in unlisted securities. The new rule provides that securities of mineral exploration companies may be the subject of dealings without the need for specific permission for each deal, providing that the company concerned has met certain requirements, namely:

- that it has nominated a broker to liaise with The Stock Exchange and undertakes to communicate all price-sensitive information to the market through this channel;
- that it has the intention of applying for a listing when it is able to meet the requirements;
- that it will not take any action to give the impression that it is a listed company.

This eminently practical approach has created what is, in effect, an over-the-counter market and must certainly be welcomed by the smaller companies as a step towards full listing.

In general terms, a company can raise new money more easily and more cheaply when the subscriber knows that there is in existence a secondary market where, should his circumstances require it, he can dispose of all or part of his holding.

New York market requirements

An American company seeking to raise funds will normally make contact with an underwriter — who may be an investment banker or a dealer — who will assess the prospects of finding subscribers for the amount and type of security envisaged. The underwriter may act alone or, if the issue is of sufficient size, form an underwriting group to acquire the securities from the company or stockholders, as the case may be, at a discount from the selling price to the public.

In the meantime, a Registration Statement will be in course of preparation by the underwriter and his counsel, the company and its counsel and various accountants, including an independent public accountant, for submission to the SEC. The Registration Statement will include a full corporate and financial history of the company and much else, and will be the subject of discussion between the principals and the SEC before it is approved. A substantial part of the document will eventually form the prospectus on which the securities will be sold. The sale of the securities will be handled by a selling group, the members of which will receive a reallowance from the price at which the securities are sold to the public as their remuneration.

With companies new to the investing public, dealings will normally take place in the Over-the-Counter market, although it is possible to indicate the intention to seek a listing on the American Stock Exchange and, if listing requirements can be met, for the listing to become effective soon after the issue has taken place.

The Securities and Exchange Commission has already been mentioned, but it is worth stressing again its very powerful position in all aspects of security markets. The Securities Act of 1933, which is enforced by the SEC, requires the registration of all new securities at least twenty days prior to public offering, in order to allow sufficient time for a thorough examination of all documents. The SEC does not, of course, pass judgment on the securities as investments but it does ensure that full and accurate particulars are available from responsible individuals to enable prospective investors to form reasoned judgments.

Cost of new issues

This item will include such things as underwriting commission, legal, accountancy and other professional fees, printing and advertising, and stamp duties. It is difficult to indicate cost levels but assuming the company has a listing these are likely to range from, say, 1.5 per cent of money raised for a simple domestic non-underwritten 'rights' issue in Britain to 5 per cent or more for a fully-underwritten major issue by prospectus and advertising, and possibly, bearing in mind that some costs, e.g. printing, do not vary much with amount of money raised, into double figures (in percentage terms) for a small issue in the United States. All these factors must be weighed, one with another, before deciding where to undertake the fund-raising.

In Conclusion

To the extent that enterprise and capital are in the same place there is obviously no need for the intervention of any third party. In practice, and ignoring very small scale businesses, this situation only applies to the extent that companies are self-financing. General growth and development of companies, complicated in recent times by the problems of inflation, requires that they have access to sources of money. In a small number of cases the access is direct, but the majority of fund-raising activities require an intermediary, which may be a clearing bank or merchant bank, a specialist organisation of some kind or a broker member of The Stock

Exchange. The unique aspect of the last-named, and hence of the organisation of which he is a member, is its dual role; in the primary market of raising new capital, and in the secondary market of subsequent dealings, over the whole range of equity and prior charge funds.

Further Reading

H. D Berman, *The Stock Exchange*, (Pitman, London 1971)

Investors Chronicle, *Beginners Please*, (London 1976)

Israels & Duff (Editors) *When Corporations Go Public* (Practising Law Institute, New York, 1970)

McRae & Cairncross, *Capital City: London as a Financial Centre* (Eyre Methuen, London, 1974)

G. Scott Hutchinson (Editor), *Why When & How to go Public*, (Presidents Publishing House, New York, 1970).

Appendix

The London Stock Exchange requirements relating to mineral exploration companies

The following is chapter 7 of *Admission of Securities to Listing* published by The Stock Exchange and is reproduced here by kind permission of the Council of The Stock Exchange. It describes the requirements applied by the Council in any application made to list securities of mineral exploration companies on The Stock Exchange. The various references in the following paragraphs to 'Schedules', 'Chapters', etc. refer to other sections of the publication *Admission of Securities to Listing.*

Scope of the requirements
1. The following requirements apply to companies whose activities (whether directly or through a subsidiary company) include exploration for and production of natural resources consisting of substances such as metal ores, mineral oils, natural gases, or solid fuels.

   An application for listing from a company whose current activities consist solely of exploration will not normally be considered, unless the company is able to establish:
   (a) the existence of adequate reserves of natural resources, which must be substantiated by the opinion of an expert, in a defined area over which the company has exploration and exploitation rights, and
   (b) an estimate of the capital cost of bringing the company into a productive position, and
   (c) an estimate of the time and working capital required to bring the company into a position to earn revenue.

Companies having no securities listed
2. In the case of a company seeking a listing for all or part of its securities but none of whose capital is already listed, the prospectus must comply with the requirements of Schedule II Part D in addition to the requirements of Schedule II Part A.

Companies having listed securities
3. Where an existing listed company proposes to include in its business exploration for natural resources either as an extension to or change of its existing activites, a circular, which must comply with Parts B and D of Schedule II, will be required to be sent to shareholders in either of the following circumstances:

(References herein to the company are to the company, its subsidiaries and principal associated companies taken together as a group.)

(a) Where the proposal involves a transaction falling within 'Class I', as defined in paragraph 5 of Chapter 4, or

(b) Where the proposal involves a transaction which might reasonably be expected to result in either the diversion of 25% or more of the net assets of the company to exploration for natural resources or the contribution from such exploration of 25% or more to the pre-tax trading result of the company. Any such transaction should be conditional on approval by the shareholders in general meeting.

In assessing the extent of diversification or the amount of contribution to the pre-tax trading result, account should be taken of any associated transactions or loans effected or intended and of any contingent liabilities or commitments.

Technical adviser

4. A company, whose activities include or are to include exploration for natural resources to a material degree, must have available to it the technical advice of a person who has had appropriate experience in the type of exploration activity undertaken or proposed to be undertaken by the company.

5. Any prospectus must contain a report by a technical adviser, as required by Schedule II Part D, which must be made up to a date not more than six months prior to publication of the prospectus. Such a report must state the name, address, professional qualifications and any relevant experience of the adviser and the date of his report.

6. A statement made in any prospectus as to the existence of natural resources must be substantiated by the professional adviser from his own knowledge and supported by details of drilling results, analyses or other evidence.

   If important evidence which must remain confidential for legal or other valid reasons has to be excluded from a prospectus, or technical adviser's report accompanying it, the company must allow an independent consultant, mutually approved, to verify to the Department in confidence the importance of such evidence.

7. Where any fee or other remuneration or consideration paid to any director, officer, technical adviser or promoter is to be paid or given otherwise than in cash, the Council reserve the right to reject any application for listing for the company's securities. In cases of doubt the Department should be consulted as soon as possible.

13

Eurocurrency Markets

The term 'Eurocurrency' is sometimes used in a context which suggests it is a special kind of currency; this is not so. For instance eurodollars are simply US dollars on deposit for the account of a non-resident of the USA. They represent the same currency and have the same monetary value as 'domestic' US dollars. The beneficial ownership of the deposit is the critical point — a dollar deposit transferred from the books of a bank in Chicago to those of a bank in London becomes a eurodollar deposit automatically, and it ceases to be such when it is transferred back to the books of a bank in the USA. This is so even if the dollar deposit remains physically located with a bank in the USA.

When a London bank uses the dollars deposited in its account to make loans in dollars, the funds so employed become eurodollar loans. Thus, we have eurodollar deposits and eurodollar loans.

The principle applies in precisely the same manner to other currencies of which the most common are Deutsche marks, Swiss francs and Dutch guilders, i.e. eurocurrency deposits are deposits denominated in a currency other than that of the country in which they are on deposit.

In addition to the eurocurrency deposit and loan market, which is primarily short and medium term, there is also a eurobond market which is a source of medium and longer term finance. However, it is crucial to note that the banks use their eurocurrency deposits to make eurocurrency loans, while it is the banks' clients who provide the funds for eurobonds. Together these markets and their complementary activities in floating rate notes and private placements comprise the euromarkets.

The prefix 'euro' may be misleading. Although eurocurrencies are primarily deposited in American banks in Europe, deposits are now also held in other countries, particularly OPEC countries. Similarly places like Nassau and Bahrain have in recent years increased in importance as trading centres; for reasons of convenience these are also frequently referred to as eurocurrency and trading centres.

Background and development

The eurocurrency market developed as a result of a number of factors arising after the Second World War. These included the very substantial inflow of dollars to certain American allies as part of the Marshall Aid plan. Other attractive capital investment opportunities in Europe for American investors appeared during the 1950s and early 1960s. The flow was further accelerated by the impetus of interest rate differentials between the USA and Europe.

It may well be that the first eurodollar deposits were placed in Paris by the State Bank of the USSR which, in the early 1950s, decided to bank some of its US dollars with institutions outside the USA. Other banks began to use the system, particularly as eurodollar rates exceeded those available in the USA. The market developed steadily over the next decade, spurred on in 1957 by British Government restrictions on lending sterling overseas and also by the practice of many European banks of satisfying domestic credit needs through the eurocurrency market.

Both in the USA and Europe there arose factors acting in favour of an enlarged euromarket. In the USA both 'Regulation Q' (a Federal Reserve Bank regulation limiting rates which could be paid on certain classes of deposit) and a chronic balance of payments deficit during the 1960s led to significant increases in both the demand for, and supply of, eurocurrencies, mainly eurodollars as illustrated by the data in table 24. Interest Equalization Tax was another factor, though more applicable to the eurobond market (discussed later).

Table 24: Estimates of size of eurocurrency markets*

expressed in US$ billions

End Period (Year)	Total (Net)	Percentage Eurodollars
1964	12	79
1966	18	81
1968	34	80
1970	65	81
1972	110	78
1974	215	77
1976 (June)	275	79

Source: World Financial Markets — January, 1977
*Interbank transactions are excluded to avoid the risk of double counting.

European central banks took advantage of the high euromarket rates for their own deposits and became important providers of funds. They felt that some control was desirable over supranational borrowers. This resulted in certain regulatory powers usually in the form of stricter reserve requirements.

London and eurocurrency market

Moscow Narodny Bank was active in dealing in the early euromarket at the beginning of the 1950s in London. Other banks became involved and London rapidly proceeded to overtake Paris as the principal centre and has remained so ever since. Table 25 illustrates the part played by London during the 1970s as a percentage of the total eurocurrency market.

Table 25: London's share of eurocurrency markets

December, 1971	46%
December, 1972	46%
December, 1973	47%
December, 1974	50%
December, 1975	50%

Source: BIS Annual Report

The structure of the eurocurrency market in London is illustrated by the 1976 data given in table 26. This shows that different types of banks have the same share in the market for both liabilities and claims (Section B of table 26).

American banks as a group are the most active in the market, and, together with the other overseas banks, they control more than three quarters of the eurocurrency transactions conducted through London. The UK participates in the market mainly through consortium, merchant and clearing banks, as shown in table 26B.

An indication of London's international position is provided by the fact that 74 per cent of British banks' customers' liabilities and 68 per cent of their claims are, respectively, to and on behalf of non-residents. The figures also indicate another feature of the London market; namely that it is largely an interbank market as 86 per cent of British banks' liabilities and 73 per cent of their claims are, respectively, to and on behalf of other banks. Other depositors, normally governments, commercial and industrial corporations, also participate although mainly via banks.

The maturity analysis of liabilities and claims of British banks in foreign currencies shows that only 6.7 per cent of all liabilities are over one year's maturity while the claims for the same maturity represent 23.8 per cent. Thus a task of London markets is to convert short term deposits into longer term loans.

Basis of eurocurrency transactions

The eurocurrency market involves a chain of deposits — of borrowers and lenders — and not of buyers and sellers. One does not buy or sell eurocurrencies.

Table 26: Structure of eurocurrency market in London

(A) Maturity analysis of foreign currency liabilities and claims of UK banks

	Liabilities	Claims
Less than 8 days	.20.4%	16.2%
8 days to less than 3 months	46.9%	37.5%
3 months to less than 1 year	26.0%	22.5%
1 year to less than 3 years	4.2%	8.6%
3 years and over	2.5%	15.2%
	100.0%	100.0%
Total (Millions of dollars)	185,675	185,997

(B) Foreign currency liabilities and claims of UK banks by type of bank

	Liabilities	Claims
UK Clearing Banks	5%	5%
Accepting Houses	3%	3%
Other UK Banks	10%	10%
American Banks	37%	37%
Other Overseas Banks	39%	38%
Consortium Banks	6%	7%
	100%	100%
Total (Millions of dollars)	185,675	185,997

(C) Foreign currency liabilities and claims of UK banks by customer

	Liabilities to	Claims on
UK Interbank Market	23%	23%
Other UK Residents	3%	9%
Banks abroad	63%	50%
Other Non-Residents	11%	18%
	100%	100%
Total (Millions of dollars)	185,675	185,997

Source: *Bank of England Quarterly Review,* December 1976

In normal domestic banking an individual, corporation or financial institution, for instance in the USA, owning a dollar deposit would normally keep it in a time or demand deposit account in an American bank. This bank would employ such funds in its operations. Until the deposit is withdrawn, control over its use lies essentially with the American bank. Similarly, an owner of a French franc deposit would keep it in an account with a French bank.

Eurocurrency operations differ from this pattern in that there is a changing control over the deposit and thus over the use to which the funds are put. As an example, a foreign corporation may deposit dollars in an account with a London bank; the London bank may deposit these dollars in turn with, say, an Italian bank, which may repeat the process or may lend the dollars to a commercial or industrial client. As the deposit moves along the chain, the control over the use of the funds shifts from one link in the chain to the next.

For a market to exist, there must be both sources of, and uses for, eurocurrencies, or, to put it another way, supply of, and demand for, them. Table 27 shows the estimated sources and uses of eurocurrency funds mainly in terms of country of origin and/or use.

Table 27: Estimated sources and uses of eurocurrency funds

expressed in percentages

	Uses		Sources	
	Dec. 73	Dec. 75	Dec. 73	Dec. 75
European area	37.2	30.7	38.6	38.8
Other Group 10 countries*	19.8	18.0	14.6	11.6
Other developed countries	11.1	12.6	13.4	9.6
Eastern Europe	5.6	7.6	2.8	2.5
Offshore banking centres**	14.2	17.3	9.5	10.7
Oil exporting banking centres	} 10.8	2.6	} 18.6	16.9
Developing countries		9.5		7.9
Unallocated	1.3	1.7	2.5	2.0
	100.0	100.0	100.0	100.0

Source: BIS Annual Report
 *USA, Canada and Japan
 **Principally Nassau, Cayman Islands, Bahrain and Curacao

Sources of funds

These are mainly derived from balance of payments deficits, from multinational corporations and from OPEC countries.

Balance of payments deficits of the USA and other countries have led to a massive accumulation of dollars outside the USA, especially in Western Europe and Japan. Interest rates paid for deposits in eurodollars have often exceeded rates on US Government paper, the excess having flowed into the euromarkets, either indirectly through the sale of the dollars to banks for local currency, or directly, as central bank deposits.

Many multinational corporations have accumulated substantial dollar and other currency holdings. Such holdings are often left as euromarket deposits, especially during the period when 'Regulation Q' discouraged American multinationals from remitting dollars back to the USA.

The fastest growing source of eurocurrencies is the revenue from OPEC oil exports. The increase in the estimated total size of the euromarkets in recent years emphasises this point. This source will assume greater importance, certainly until the mid 1980s.

Uses of funds

Over the years the amount of money needed by both governments and corporations has increased beyond the capacity of most domestic capital markets — except perhaps the American market, which has effectively been closed to most foreign borrowers. This clearly has special relevance to the large oil companies, whose requirements are enormous even by the standards of capital intensive industry. Nowadays, many projects, such as oil production in the North Sea, commonly run into hundreds of millions of dollars. The eurocurrency market is now practically the sole source of funds in such amounts, excluding the American market and internally generated funds of the individual company. The euromarkets are also seen as flexible means of raising the capital necessary for development.

The interbank market

A major part of eurocurrency activity is comprised of re-deposits among international banks active in the market. This interbank market has both borrowing and lending aspects.

A bank accepts eurocurrency deposits from individuals and corporations, but also from other prime banks. A bank will accept deposits and on-lend them for periods varying from overnight to five years. Longer periods are more unusual and harder to obtain. The normal period for which a bank takes a deposit is three or six months. Such deposits are called interbank deposits, and the interest rate payable on them is known as the London Interbank Offered Rate (LIBOR) for the eurocurrency in question for the stated amount and period. There is a bid and an offered rate.

Particularly in the major centre, London, there is a sophisticated system run either directly between banks or via moneybrokers, which ensures that the market is extremely quick to react, enabling an efficient interbank market to be operated. Such interbank deposits may either be used for a bank's own operations or on-lent perhaps several times before reaching a non-bank commercial borrower.

Because the interbank deposit market is almost perfect, the LIBOR for a given borrower at any time varies only minutely, if at all, between different banks (whether London based or not).

Short term interest rates

Balance of payments deficits, coupled with relative inflation rates, play a part in determining the relationship between eurocurrency rates as well as their absolute levels. Government/central bank regulations can cause rates to fall and even go negative in the case of strong currencies, such as the Swiss franc.

A meaningful source of information about the structure of interest rates is provided by comparing domestic rates and eurocurrency rates. This relationship explains why the euromarket exists and thrives. This shows why companies, governments and other institutions, deposit and borrow dollars in London, for instance, rather than in New York.

Interest rates on eurodollar deposits have occasionally fluctuated widely, but are principally determined by the interest rate structure in the USA. Slack loan demand experienced in the USA has a depressant effect on domestic rates, as was seen in the recession of 1974-75 and this was partly reflected in the eurodollar rate. Chart 1 in figure 1 shows that the movements of interest paid on federal funds have overlapped the line of overnight eurodollar rate for the last two years. This suggests that the same factors affect US dollar domestic rates and eurodollar rates. The trend of both rates would be so determined by the expectations on US rates, by the liquidity of American banks and by the behaviour of the US monetary authorities. Although peaks and troughs are the same for both rates, deposit rates in the eurocurrency market are always higher than the corresponding ones in the USA as shown in chart 2 in figure 1.

On the loan side, a straight comparison between the prime rate (the rate that top customers have to pay to American commercial banks) and eurodollar rate can be misleading. A potential borrower is concerned with the overall cost of borrowing. If compensating balances, always requested by American banks, are added to the prime rate, it is apparent that the cost of a eurodollar loan (LIBOR plus spread) has always been cheaper than a dollar domestic commercial credit (prime rate plus compensating balances) — as shown in chart 3 in figure 1.

The relatively low intermediary charge of the euromarket is a comparative advantage that can explain the existence of the market itself. In fact, this is not wholly an institutional market but is based on the mutual convenience of both depositors and borrowers. It will survive only as long as this convenience lasts in the future.

Alternative methods of borrowing eurocurrencies

The recognised techniques used to lend eurocurrencies to commercial borrowers can be broadly divided into two categories:

- eurocredits — fixed or more usually floating rate syndicated loans;
- eurobonds — fixed, and floating rate notes.

There are specific distinctions between these categories. Eurobonds are generally for longer periods than eurocredits. They are also normally at a fixed interest rate. Another difference is that eurobonds are marketable securities.

Assuming that the demand for syndicated loans and the supply of bonds are explained by their respective cost, these will be determined by the term structure of interest rates. An investor purchases a long term asset when its yield is higher than his expectation for the average

Figure 1: US and eurodollar interest rates*
Chart 1

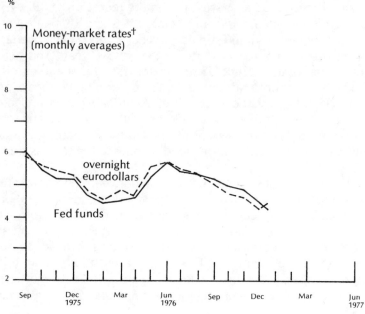

Chart 2

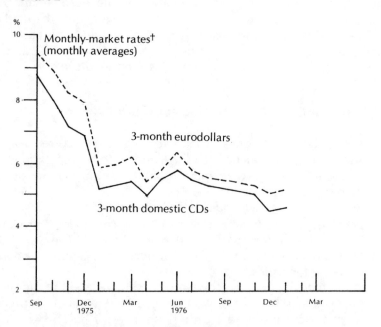

Source: World Financial Markets, January 1977.

*excluded effect of Regulation M reserve requirements
ᴸLatest plotting average of three weeks ended January 21, 1977.

Chart 3

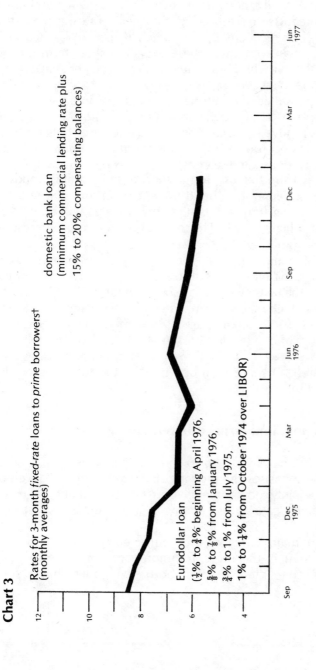

Rates for 3-month *fixed-rate* loans to *prime* borrowers†
(monthly averages)

domestic bank loan
(minimum commercial lending rate plus
15% to 20% compensating balances)

Eurodollar loan
($\frac{1}{2}$% to $\frac{3}{4}$% beginning April 1976,
$\frac{5}{8}$% to $\frac{7}{8}$% from January 1976,
$\frac{3}{4}$% to 1% from July 1975,
1% to $1\frac{1}{4}$% from October 1974 over LIBOR)

12

10

8

6

4

Sep | Dec 1975 | Mar | Jun 1976 | Sep | Dec | Mar | Jun 1977

†latest plotting average of three weeks ended January 21, 1977

Source: World Financial Markets January 1977 reproduced by kind permission of the
Morgan Guaranty Trust Company.

of future short term rates during the life of the asset, plus a premium for the 'liquidity lost'.

An example can be helpful — an investor must make up his mind about a seven-year bond with a 7 per cent yield. He thinks one per cent represents his illiquidity premium. Then he will purchase the bond if he forecasts the following short-term rates for the forthcoming seven years respectively: 5%, 5%, 7%, 8%, 6%, 4% and 4%. This is because the average of these expected short term rates (5.6 per cent) together with his illiquidity premium (one per cent) still total less than the 7 per cent offered by the bond.

This could be called the 'normal' structure of interest rates. It is important to point out that the structure does not change even if short term rates move up to and beyond the level of long term rates during the third and fourth years. This is true assuming no change occurs in expectations. On the contrary , if this happens, our investor will shift money from the long to the short term by selling bonds. The sale will cause a decrease in bond price, that in turn will result in a higher yield on long term assets. Eventually, rates will reach a new equilibrium at a higher level. The 'normal' structure of interest rates will then be re-established.

The above dynamic process is, of course, an oversimplification. Nevertheless it is extremely useful to explain the vicissitudes of euro-market rates in the last few years as shown in table 28. 1973 and 1974 were years of high short term interest rates and bond issues accounted only for a small part of the medium and long term eurocurrency markets (table 29). In 1975 and 1976, when the expectations were again for a lower level of interest rates, bond issues accounted for more than one third of the total eurocurrency market (table 29).

The short and medium term eurocredit market

A geographical analysis of borrowers in this market is shown in table 30 and emphasises the growing significance of the developing countries — mainly governments borrowing to meet balance of payments deficits caused by, *inter alia*, the increased cost of oil imports.

Of the two main categories of eurocredits noted earlier, the floating rate loan is the most common form. However, with the exception of the interest rate, both categories have similar characteristics.

Table 28: Eurocurrency markets — selected rates*

expressed in percentages

	Dec. 1973	Dec. 1974	Dec. 1975	June 1976	Dec. 1976
Deposits					
USA (3 months CD)	9.25	9.25	5.50	5.75	4.70
Eurodollars (Prime banks' bid rate for 3 months deposit in London)	10.13	10.19	5.81	6.00	5.00
Bond Yields					
European companies (Eurobonds denominated in $)	8.56	11.23	9.62	9.27	8.56
Governments (Eurobonds denominated in $)	8.69	10.08	9.30	9.36	8.90
Loans					
Prime rate USA + compensating balances (20%)	11.70	12.30	8.70	8.70	7.20
Eurodollar + spread (Average rate for 3 months to prime borrowers)	10.75	11.32	6.69	6.63	5.63

*At or near end of month
Source: Morgan Guaranty Trust

Table 29: Medium and long term eurocurrency market

expressed in billions of dollars

	1973	1974	1975	1976
Eurocredits	21.8	29.2	20.9	28.9
Eurobond issues	4.2	2.2	8.6	14.0
Total credit facilities	26.0	31.4	29.5	42.9
Eurocredits	84%	93%	70%	67%
Eurobond issues	16%	7%	30%	33%

Source: Morgan Guaranty Trust and IMF survey

Table 30: Medium and long term eurocredits

expressed in percentages	1973	1974	1975	1976
Industrial countries	63	71	34	37
Developing countries	21	21	39	39
OPEC countries	12	4	14	14
Communist countries	4	4	13	10
	100	100	100	100
Total (billions of dollars)	21.8	29.2	20.9	28.9

Source: Morgan Guaranty Trust

Amount and currency

The size of the loan may be anything from $250,000 to over $1,000 million (the largest to date being that for $2,500 million arranged in 1974 for the UK government). In other words, the capacity of the market can encompass the needs of most borrowers, including the oil companies. The currency of the loan will depend on the borrower's needs and, to some extent, the availability of currencies; the dollar is still the predominant currency, although loans in Deutsche marks, Swiss francs and others are regularly arranged. Increasingly, we see multi-currency loans, in which the borrower has the right (subject to availability) to switch from one currency to another. This flexibility enables the borrower to match his loan with the currency of his cash flow to meet interest payments and principal repayments. It also allows him to take advantage of the lowest LIBOR after considering foreign exchange risks.

Period of the loan

The period of the loan can vary from, say, one year up to ten years but the most common period now seems to be from five to seven years. Events such as the collapse of a number of banks in the USA and West Germany in 1974 have profound implications and effects on the confidence of the international banking community and their willingness to extend loans with longer maturities.

The drawdown and repayment periods vary in accordance with the borrower's particular needs. Repayment schedules are most commonly begun after about half the total life of the loan. If, in a five-year loan, the period before repayments begin (known as the grace period) is 30 months, repayment would typically be effected by six payments, each being one sixth of the original principal amount, made at the end of months 30, 36, 42, 48, 54 and 60. Full repayment at maturity (known as bullet repayment) is less attractive to banks and could influence the lender towards reducing the final maturity in order to bring the average life of the loan to a level comparable with that for other loans having a longer final maturity and a repayment schedule. A commitment fee is paid — 0.5 per cent per annum is normal, but subject to negotiation — on undrawn portions, while prepayments, in advance of the agreed schedule, are allowed, sometimes subject to payment of a penalty.

Costs, risks and associated guarantees

The interest charged on floating rate loans to borrowers is based on a margin above LIBOR for the currency and period agreed. The term of the deposit used as the reference may be one, three, six or twelve months, and at the end of each respective period interest for the next period is calculated usually at the same fixed margin above the new prevailing LIBOR. The most frequently used period is six months, and after each such period the rate is re-set for the next period. This is known as 'rolling over' the loan. The margin, turn or spread, i.e. the difference between the cost of funds to the lending bank and the interest charged to the borrower, varies a good deal. The finest rates in mid-1977 were around 0.75 per cent per annum above LIBOR for certain quality borrowers and up to 3 per cent above LIBOR on the higher risk loans. The factors governing the size of this spread are many and include the credit standing of the borrower, the amount and the period of the loan, the arrangements for its repayment, the nature of guarantees provided, whether or not it is otherwise secured and, especially, the 'country risk'. Competition between lenders can also be a factor. For instance, spreads narrowed during the course of 1972 and 1973 and again in 1976 and early 1977, often down to levels where many banks refused to participate, not because they doubted the credit, but because the return on the capital allocated would be unacceptably low.

Country risk is probably the first and vital consideration. Historical evidence of political and social stability, previous experience with borrowers in the area, past and estimated future economic performance, including the level and service of external debt and the fraction of hard currency exports required to meet interest and principal payments due, all combine to form a picture of the risk pattern to be expected in lending to the country over the life of the loan. Company analysis covers the normal range of credit analysis prerequisites. These would include examination under categories of historical, present and future forecasts of market, managerial and financial performance.

Guarantees are an important further aspect affecting the general terms of borrowing. It is usual for companies to guarantee subsidiaries or affiliates taking balance sheet loans though not, by definition, loans on a 'project' basis (see below). If a company can arrange for its local bank, which knows it on a day-to-day basis, to guarantee a loan made by a foreign bank, it is normally found that the reduction achieved in the cost of the loan is greater than the guarantee fee payable to the borrower's local bank. Suffice it to say that some lenders prefer a bank guarantee and accept the lower yield while others, possibly knowing the borrower's geographical area well, would not require a bank guarantee.

Security offered to banks varies considerably. At one end of the scale is the fully compensating deposit, though perhaps in a different currency from that of the loan. At the other end, the loan may be totally unsecured except for a 'side letter' providing 'comfort' to the lender. In the latter instance restrictions on other borrowings or a 'negative pledge' to refrain from securing assets elsewhere during the life of the loan would often be necessary in lieu of security.

Organisation of the loan — the lenders
Lenders in this market are almost exclusively banks. In any single loan the number of participating banks will depend on the size of the loan and the size of the lending bank. Some of the largest banks can lend up to $30 million or $50 million in one transaction and one such bank may therefore be the only participant in a loan.

It is more normal, however, to syndicate a loan of $10 million or more. The structure is to have the originating bank as the manager of the loan, and this bank may invite one or two other banks to co-manage. The function of the manager(s) is to bring other banks into the loan as participants. The manager is normally responsible to the syndicate of banks for ensuring that the borrower performs throughout the period of the loan in accordance with the terms and conditions of the loan agreement, though of course the participants have recourse only to the borrower. The manager charges the borrower a once-and-for-all fee, which is a percentage of the total amount of the loan (0.5 per cent to 1.5 per cent) and some of which is distributed to the other participating banks.

The loan agreement
The evidence of the loan is a legal agreement, signed by all the participating banks as well as by the borrower and by any guarantors. The form of the loan agreement has become fairly standard and covers details of not only the loan terms (the most important of which have been discussed above), but also the allocation of responsibilities for the multitude of events that may occur during the life of the agreement and which affect the loan.

One of the more usual clauses in a loan agreement is that relative to tax. In most cases such a reference states that taxes are for the account of the borrower, since banks do not wish to receive interest net of withholding tax (unless there is agreement with the bank, in which case there will be an appropriate indemnity from the borrower covering the bank, should the latter not be able to offset the tax deduction against its own tax liability). More frequently, instead of interest payable being increased to include withholding tax the loan agreement may provide for an amount equal to the tax withheld to be placed on deposit with the lender. This 'compensating deposit' is refunded to the borrower after the lender has received a double tax credit for the withholding tax. As shown in table 31 the level and treatment of withholding tax varies widely according to the country of domicile of the borrower.

Another important clause in a loan agreement is that relating to default. Default is deemed to take place not only when interest or principal is not paid or repaid respectively on the due date, but also for example in the event of the borrower's making false or inaccurate claims and representations, if the borrower declares a moratorium on the payment of indebtedness, or if the borrower enters into voluntary or involuntary liquidation. A default clause would also include within its terms of reference the obligations of the guarantors. Conditions similar to those imposed on the borrower are often brought to bear also on the guarantor, and in fact the theoretical possibility exists that if the guarantor were to become bankrupt while the borrower itself remained solvent, there would then be an event of default.

Table 31: Withholding tax ranges for certain major borrowers in medium term euromarket 1976

	Rate
Brazil	25%
Mexico	10% — 42%
Spain	1.2% — 24%
UK	0 — 35%
Venezuela	10% (5% increase planned)
Argentina	11.5%
Philippines	35%
Iran	15%
USA	0 — 30%
Norway	0 — 25% (not on interest)
South Africa	10%
Canada	0 — 25%
Denmark	0 — 30% (not on interest)
France	0 — 33⅓%
Sweden	0 — 30%

Source: HMSO
Note: The rates applicable for each country are subject to a host of individual regulations. For certain transactions the indicated rates are reduced or waived altogether. Withholding tax rates are always subject to change.

One of the shortest clauses in a loan agreement is also one of the most significant; it is that relating to the law and jurisdiction by which an agreement will be governed. The vast majority of loan agreements for eurocurrency loans are governed by, and construed in accordance with, the laws of England. The main reason for this is that English law, like the English language, is respected, recognised and accepted throughout most of the world. English law is preferred to USA law, because the legal process is less time consuming and less costly in the event of legal action being necessary.

In a large syndicated loan there are also undertakings from the managers to make sure each member of the syndicate is aware of any new developments arising or changes proposed in the terms of the loan and that they always receive relevant information pertaining to both the borrower and the guarantor.

In general, the loan agreement and its many long and often complex clauses would seem to be of material benefit to the lender alone. In fact, there are many ways in which the loan agreement may be beneficial to the interests of the borrower as well. It obviously enables both the lender and the borrower (but especially the latter) to know where each stands in relation to the other, to know at a glance what the regulations are and what is or is not permissible. It enables the borrower to know that the funds he will eventually receive are sound, that the interest rates quoted to him will be based on the market rate and that the status of the lender is good. If a large syndicate of lenders is involved, the borrower will certainly wish to be assured of the worth of each of them. With regard to the governing law, it is in fact often the borrower, as much as the lender, who will insist on English law — the borrower knows full well that notions of impartiality may not be as strong in other legal frameworks.

Fixed rate loans

During the mid 1970s the market for fixed as opposed to floating rate loans diminished significantly. The principle however is very similar, involving a London interbank offered rate for deposits of the required term — sometimes as long as five years. To this is added the margin or spread making up the total interest cost to the borrower. The documentation is entirely analogous to that required for floating rate transactions though it was not unusual for these loans to be made on the basis of repayment only at maturity. From the fixed rate aspect, the loan resembled a eurobond, with the important difference that it was not openly marketable.

The longer term eurobond market

Eurobonds[1] may be defined as securities marketed internationally in an internationally acceptable currency, which need not be that of either the borrower or the investors. The fundamental distinction between a eurocredit and a eurobond is that the latter is openly marketable and quoted in a 'secondary' market. Until the early 1960s, there was no means by which eurocurrencies could be borrowed for long term periods. To meet this demand for long term finance and as a result of the increasing supply of eurocurrencies available coupled with attractive interest rates, the eurobond market rapidly developed after about 1963. Like the eurocurrency credit markets, the US dollar is the predominant currency. In 1963, the US Government established guidelines for direct investment abroad and introduced the Interest Equalization Tax. This tax made US residents reluctant to purchase foreign bond issues on US stock exchanges, thereby confining the market to non-US resident purchasers. A further spur to the bond market's development was that European industry, which had for long relied on internally generated funds to finance its long term growth, found that such funds had become inadequate. Industrial development had reached the point where additional sources of long term finance were needed if future growth was not to be impaired.

Eurobonds developed in response to this need for long term capital. In May, 1963, the first issue recognised as a eurobond was raised in London. It soon became apparent that a substantial international market could flourish in London, and the City quickly became the leading centre for the issue of eurobonds (although ironically UK investors in eurobonds are very rare because of Bank of England curbs on foreign currency investments). Certain institutions, often merchant banks, have come to specialise in managing eurobond issues.

The size of the market may be judged from these figures: in 1976, borrowers raised a record total of $14,000 million after $8,600 million in 1975, $2,966 million in 1970, $1,142 million in 1966 and $164 million in 1963.

The relative importance of the various categories of borrowers is shown in table 32.

Characteristics of bonds

Eurobonds, like large roll-over credits, are issued by governments, government creatures and large industrial concerns of international standing, especially the multinational companies. Into this category, naturally, fall the oil companies, and many well-known oil companies have made demands on the eurobond market. Factors such as the

Table 32: Eurobond issues

expressed in percentages by category of borrower

	1973	1974	1975	1976
US companies	21	5	3	3
Companies outside USA	31	30	34	37
State enterprises	22	25	36	28
Governments	16	23	20	16
International organisations	10	17	7	16
	100	100	100	100
Total (millions of dollars)	4,193	2,134	8,567	14,036
US dollar issues as a percentage of total issues	58.4%	46.7%	43.5%	63.6%

Source: Morgan Guaranty Trust

reputation and creditworthiness of the borrower and whether it is well-known are more significant to the success of a eurobond issue than in the case of a eurocredit loan.

*Size of Issue:*or amount to be borrowed, is in general considerably smaller than amounts raised in eurocurrency credits. At any point in time the amounts which can be raised depend upon investor confidence and availability of funds. Typical amounts have been $50 million to $75 million and often much smaller with an issue of $100 million or more being rare. The largest issue to date is the emergency $500 million raised for the EEC in 1976. This was exceptional. Still, many analysts are forecasting an increasing number of larger sized issues for the future and this is clearly a useful development for the oil industry.

Period: eurobonds have had maturities ranging from 5 to 20 years, although 5 to 10 years have been most common. To make an issue more attractive a sinking fund, or a purchase fund, might be introduced. Some bond issues have been convertible into shares, provided of course the borrower is a quoted company.

Generally, there is only one drawdown of funds and a set repayment schedule exists, which cannot be varied by the borrower without incurring a penalty.

Coupon or rate of interest: this depends on factors such as the currency, amount, period, repayment arrangements, market conditions, short term interest rates and the reputation and credit standing of the borrower. Through their knowledge of the supply of funds and prevailing interest rates, the issue manager or managers can gauge how much can be raised, what the coupon should be and whether the issue price of the bonds should be at or below (or even above) par.

Mechanism of the primary market

Eurobonds resemble ordinary bonds (such as US government bonds or UK gilt-edged) in the way they function more than they differ from them. They are quoted securities on sale to the investing public; they bear a coupon paying the advertised rate of interest on the nominal amount to the holder; and at maturity they are redeemable at par.

Like eurocredits, new eurobond issues are syndicated. The method is a three-tier system of banks, made up of managing underwriters, sub-underwriters and a further group, known as the selling group, though all three have responsibilities for marketing the issue. The purchasers are by and large institutional investors, funds and individuals, who are permitted under the laws of their country of incorporation to hold foreign currency securities.

Two patterns of underwriting are used in this market — normally referred to as the British (also widely used by continental banks) and the American method. Using the British method, the managing underwriters act as principals and legally purchase the bonds from the borrower. They then invite other banks to act as sub-underwriters to cut their own risk. Under the American method, the lead manager legally acts as an agent and not as principal, acting only on behalf of the other underwriters who buy the bonds from the borrower.

When a new issue is announced in a press advertisement called a 'tombstone'[2], the bonds will have already been 'placed' with the underwriters or the sellers (and the proceeds will have already been paid to the issuer). If either the underwriters or sellers cannot sell their agreed 'quota', then they fund it themselves and dispose of the bonds on the secondary market when the bonds are quoted.

The commissions normally amount to 2—2.5 per cent on the principal amount of the issue (including 0.375—0.5 per cent for management; 0.375—0.5 for underwriting and 1.25--1.5 per cent for the selling group). The commissions paid depend on the maturity of the issue and normally are as follows: five years 2 per cent; seven years 2.25 per cent and ten years and above 2.5 per cent.

Bond Contract: the evidence of a eurobond issue is to be found in a series of contracts to cover every right and obligation from the issuing of eurobonds to their eventual redemption. The detailed content of the contracts will depend upon the pattern of underwriting adopted. It will incorporate many of the general clauses referred to in connexion with eurocredit agreements, e.g. law and jurisdiction. The full documentation of an issue will be retained by the principal parties to an issue (the borrower, managing underwriter and the trustee, if any). The investors will have access to these during the course of an issue and a description of the terms and conditions of the bonds will be given in the prospectus and on the bond itself.

Secondary market

When a new issue has been subscribed and allotments have been made (this stage being known as the primary market), a secondary market in the bonds follows. In this market the new issue is added to the list of all the other existing bonds, being then subject to the usual price fluctuations that abound in all securities markets.

Trading in eurobonds is normally on an 'over the counter' basis. As noted earlier, eurobonds are bearer securities and most holders usually arrange for their bonds to be held for them in the two principal clearing houses — Euroclear in Brussels and CEDEL in Luxembourg. These institutions enable the whole settlement and delivery aspects of the eurobond market to be carried out with exceptional simplicity.

Recent developments in the eurobond market

The eurobond market fulfils a real need and could not readily be replaced. It is likely that increasing attention will be paid to the large-issue private placement bond market (where issues are not publicly subscribed and not listed on an exchange, but are placed by the issue managers). The opening months of 1977 illustrated this possibility when Mobil and Shell raised $500-million between them with private placements.

There is a growing popularity for floating rate notes, whereby the coupon is not fixed for the life of the issue but is based on the six month eurodollar deposit rates, plus a margin. Therefore, the coupon fluctuates and is fixed at six monthly intervals, in much the same way as roll-over credits are arranged.

Another discernible trend starting in 1975 was the investor's preference for medium term five to seven year issues. However, during 1977 there has started to emerge a greater degree of acceptance by investors of longer term maturities in the range of ten to fifteen years.

Finally, the first 'Asian dollar bonds' were issued in 1976, employing funds from the embryo Asian dollar market, based largely in Singapore. Analysts predict further expansion in the Far East and also possibly in other areas of the world where surplus oil revenues are to be found.

Petroleum industry involvement

An indication of the degree to which the international petroleum industry has made use of the eurocurrency markets is given by the brief data summarised in table 33. This shows that some 13 per cent of eurocredits and 5 per cent of eurobonds in 1976 were destined for the petroleum industry. Most of these loans were granted on balance sheet considerations (with necessary credit analysis) and are serviced and repaid from the company's overall cash flow.

Another category, project loans (defined in part C), rely on the revenue from a specific project or scheme for payment of interest and repayment of principal. As an example of project finance, and also to illustrate petroleum industry involvement in large-scale eurocurrency financing packages, mention may be made of the method by which British Petroleum was able to finance the development of its Forties field in the North Sea, the first major financing in this area.

Table 33: Petroleum industry involvement in eurocurrency markets*

expressed in millions of US$ equivalent

Eurobonds

	Total issues of oil industry	As a percentage of total eurobond issues
1975	916	10.8%
1976	801	4.7%

'Typical' eurobond issued by oil industry**

	Average amount`	Average maturity (years)
1975	36.7	10.8
1976	40.0	9.2

Private placements

	Total amount	Average amount	Average maturity
1976	514	57	10.2 years

Eurocurrency credits

	Total credit to oil industry	As a percentage of total eurocurrency credits
1975	3108	14.4%
1976	3635	13.4%

'Typical' loan to oil industry

	Amount	Maturity (years)	Spread in percentages
Average	81.2	5.4	1.5
1975 (Maximum)	380	10	2.25
(Minimum)	4	1	1.125
Average	79.0	6.4	1.625
1976 (Maximum)	430	10	2.375
(Minimum)	10	4	1.250

*A broad definition of oil industry has been used; therefore, drilling companies, refineries, etc., have been included.
**The average rate is impossible to calculate as bonds are issued in different currencies. Rates reflect the strength of the currency in which the bond is denominated.

The table reproduced above is reprinted by kind permission of *Eurostudy,* the review of Euromarkets edited by William F. Low.

Financing the Forties field

The Forties field is a major oil field by any standards and of great importance to the British economy. In an uncertain and hostile environment, the cost of developing such a field is very considerable, and the original estimate in 1972 put the figure at about £360 million though by 1977 it had risen to over £800 million. Although some of the original estimated cost could have been met from other sources, it was proposed that this sum (£360 million) be raised by a large syndicate of banks. Accordingly, an agreement was signed in June, 1972, to make the sum available in both sterling and US dollars (i.e. £180 million and $468 million, the latter was then equivalent to the other £180 million). This loan has since acquired considerable fame and, because its basis is similar to the 'production payment' financing used for many years in the USA, has become a classic example of financing new oil developments.

The method is as follows:
- —The banks lend to a non-profit making company, NOREX Trading Limited, which is controlled by the syndicate managers.
- —Under a Forward Oil and Gas Purchase Agreement, NOREX buys from a BP subsidiary, BP Oil Development Limited, oil (and gas) to be produced from the field for a consideration of up to the equivalent of about £360 million. The advance payments made by NOREX for such oil are available to finance the development of the field. Minimum quantities are to be delivered from the start of production but certain flexibility is retained as to how much oil BP Oil Development will supply to NOREX.
- —Under a Long Term Sale Agreement, NOREX sells delivered oil to BP Trading Limited at pre-determined prices which increase over the period of the agreement. The sale of the oil to BP Trading generates a cash flow which is applied by NOREX to service its loan from the banks. Since the oil is being paid for by NOREX in sterling and dollars, it is bought by BP Trading for the appropriate currencies so that NOREX has no exchange risk.
- —During the period before the development is successfully completed BP Oil Development reimburses NOREX in cash for the interest payments which NOREX has to pay on the loan from the banks. Subsequently, interest payments are to be met from the delivery of oil to NOREX.
- —BP indemnifies NOREX, the managers and the syndicate members against any failure by BP Oil Development and BP Trading to fulfil their obligations under the various agreements.

The syndicate agreed to accept the risk that the value of the oil in place in the Forties field might be insufficient to service the financing, while BP had underwritten the recoverability of such oil. The initial drawdown was made in the first half of 1973, and further drawings were made up to the time the development of the project was completed.

Repayments, which began as soon as production itself began, are spread over five years and must in any case be completed by the end of 1982. In fact, the proviso was made that repayments would be suspended if the oil flow were interrupted but, as we now know, production is taking place 20 per cent faster than anyone thought likely (i.e. maximum output is now expected to reach 500,000 barrels per day compared with a projected 400,000 bpd); indeed the higher forecast of production will enable BP to accelerate the scheduled repayments of the loan. It is expected that about 50 per cent of the oil produced from Forties in the five years of repayment will enable the production payments to be met.

As security for the advances, BP has charged the assets and contracts connected with the development of the field, and in the unlikely event of default by BP, the company controlled by the managers, NOREX Trading, will take control of Forties and the facilities for operating it.

Conclusion

The eurocurrency markets, since their inception, have filled a need in the field of international finance and have proved remarkably resilient in the face of adverse world market conditions. There is no evidence that their relevance or importance will change in the medium term future. However, in the longer term the following points might be considered.

At the present time, the world lacks a monetary system, in the sense that there is no official institution capable of supplying the international payment system with the required liquidity. This function has been taken over by the private banking system and primarily by the eurocurrency markets. This is supported by the nature of the borrowers; governments financing deficits in their balance of payments; developing countries additionally financing the growth of their industrial and social infrastructure and COMECON countries supporting their purchases of Western technology.

The eurocurrency markets are supranational and therefore are not regulated by any one national authority. The possibility of some form of regulation of these markets is currently being discussed. One suggestion is that the IMF might be an appropriate organisation to perform this function.

The high degree of flexibility of the eurocurrency markets particularly suits the needs of the oil industry whose borrowing requirements in general exceed those of other industries. Furthermore, their requirements are often too large for the capacity of the domestic markets with the exception of the USA. However, many non-US borrowers have limited access to that financial market owing to the strict rules prevailing there on the required 'standard' of borrowers and the regulations which govern the extent to which US financial institutions are permitted to invest in securities issued by foreign borrowers.

Finally, though it is easy to reach the conclusion that the eurocurrency markets will continue in the future, it is more difficult to forecast how these markets will grow.

References

[1] Technically, notes are those with a maturity from issue of up to five years, bonds those with a maturity of above five. Generically, they are all 'eurobonds'. Perversely, however, it is possible to find seven-year notes and five-year bonds. Practically, there is no difference whatever between the two.

[2] The purpose of advertising is to show both the managing bank's achievement in syndicating the loan successfully and the borrower's creditworthiness. Usually a line in small print appears at the top of the advertisement, saying 'This announcement appears as a matter of record only'. This is why these advertisements are called tombstones.

Appendix
Glossary of terms

Eurocredits

Agent bank: The bank assuming responsibility on behalf of a syndicate of banks for the operation of a loan and communications with the borrower.

Bullet repayment: A loan in which repayment is made in one amount at final maturity.

Co-manager: A bank ranking after the lead manager or manager in the structure of a syndicate; each co-manager will undertake to underwrite a sizable proportion which will largely be placed with other syndicate members.

Commitment fee: Fee payable by the borrower to the lender(s) in relation to funds committed by the latter but not drawn down.

Drawdown: The point when funds due to the borrower are actually taken by him for the purpose prescribed. This may take place in one sum, or in several instalments fixed in the loan agreement or may be left at the borrower's discretion.

Eurocurrencies: Bank deposits beneficially held outside a currency's country of origin; i.e., dollar deposits held in London. Likewise euro-Deutsche marks, euro-French francs.

Grace period: Period of the loan between when drawdown can take place and the begining of any repayment schedule.

LIBOR (London Interbank Offered Rate): The interest rate payable on interbank deposits for a given currency for a stated amount and period.

Loan agreement: A legal agreement which represents the evidence of a loan detailing its terms and covering the multitude of events that may occur during the life of the agreement and which affect the loan.

Manager: The bank responsible for bringing a loan to the market by offering participations to co-managers and syndicate members and for negotiating loan terms with the borrower.

Management fee: An initial fee charged by the manager to the borrower, some of which is passed on to co-managers. It typically ranges from 0.5 per cent to 1.5 per cent of the amount of the loan.

Placement memorandum: Document circulated by a manager to potential syndicate members detailing terms and conditions of a proposed borrowing. Although circulated by the manager, the contents are the explicit responsibility of the borrower.

Prepayment: Repayment of a loan made in advance of an agreed schedule, sometimes subject to a penalty.

Repayment schedule: A stipulated timetable according to which loan repayments are to be made, e.g. six equal semi-annual sums over the final three years of a five-year loan.

Roll-over: The procedure in the operation of medium term loans whereby interest rates are adjusted at regular intervals to acknowledge change in the current market rates for short term funds. This is the alternative to fixing the interest rate for the entire period of the loan.

Spread: The margin (typically 0.75 per cent to 3.0 per cent) payable by the borrower in excess of the prescribed base interest rate (usually LIBOR).

Syndicate: The entire group of banks (including manager, co-managers, agent) which collectively provide funds for a loan, each of them taking over a relatively small proportion of the total on its own account.

Tombstone: An announcement in the financial press, not unlike tombstone inscriptions in appearance, which lists the basic terms of a loan or new issue, with names of borrower and managing banks, followed in smaller print below by the names of the syndicate members.

Withholding tax: A deduction sometimes made by governments on interest paid to non-resident lender.

Eurobonds

AIBD (Association of International Bond Dealers): founded in 1969 for the establishment of uniform market practices. It now has over 350 member banks active in the issuing and secondary markets.

Asked: Price demanded by seller in the secondary market.

Bid: Price offered by purchaser in the secondary market.

Bullet: A straight-debt issue without a sinking fund with repayment taking place in full at maturity.

CEDEL: One of the market's two clearing systems, owned by several European banks.

Convertible: Fixed-interest borrowing convertible into the borrower's common stock on stipulated conditions.

Coupon: The fixed interest rate attached to a eurobond.

Eurobond: A security marketed internationally in an internationally acceptable currency, which need not be that of either the borrower or the investors.

Euroclear: One of the market's two clearing systems, provided under contract by Morgan Guaranty for over 100 banks which own it.

Floating rate notes: Securities issued on the eurocurrency market with a floating rather than a fixed rate of interest; this rate of interest is determined by reference to a formula linked to the six months eurodollar rate. Often such notes carry a minimum interest rate.

Foreign bond: A security issued by a borrower in the national capital market of another country (as distinct from a eurobond, marketed internationally). Except in New York, the flow of new foreign issues on national markets is regulated, if permitted at all.

Issue price: The price at which bonds are sold on issue, normally expressed as a percentage of the bond's face value, which is usually $1,000 in the eurobond market. Par is the equivalent of face value; a sale price of 99.5 means that the bond is sold at a discount for $995; quotations above 100 represent a premium over face value.

Lead manager: Bank responsibile for bringing a new issue to market.

Listing: Quotation on a stock exchange, normally London or Luxembourg for eurodollar issues, Frankfurt for euro-DM bonds, Luxembourg for euro-French franc issues.

Purchase fund: This is an alternative to a sinking fund provision. The borrower is obliged to purchase a stated number of bonds only if the price of the bonds is below par or the original issue price, assuming that the bonds are available for purchase.

Secondary market: Market in which bonds are traded after issue, normally in minimum lots of $10,000 or 10 bonds dealt through banks acting as market makers.

Selling group: All banks marketing a new eurobond issue.

Sinking fund: A fund provided by the borrower for the repurchase of securities during their life, the effect being to reduce the average maturity of the issue as a whole.

Underwriting group: Banks receiving 0.375 to 0.5 per cent fee for 'underwriting' a new eurobond issue, although they are seldom required to fulfil the obligation for which they have been paid. The full amount of any new eurobond issue is always underwritten by the manager, co-managers and underwriting banks, but since part is usually marketed by selling group banks who are not underwriters, the underwriters as a group need rarely take up their full commitments.

Yield: The return on a fixed-interest security based on coupon price either to its average remaining life (yield to average remaining life) or to its remaining life (yield to maturity).

Further reading

Suggestions for further reading:

Bank for International Settlements, *Annual Reports*, Basle

International Monetary Fund, *Annual Reports*, Washington

Morgan Guaranty Trust, *World Financial Markets*, New York

William F. Low (Editor), *Eurostudy,* London

Bank of England, *Quarterly Bulletin*, London

World Bank, *Borrowing in International Capital Markets*, Washington

Euromoney — (monthly magazine), London

B. S. Quinn, *The New Euromarkets*, (Macmillan, London, 1975)

G. W. McKenzie, *The Economics of the Eurocurrency System*, (Macmillan, London, 1976)

G. Bell, *The Eurodollar Market and the International Financial System*, (Macmillan, London, 1973)

P. Einzig, *The Eurodollar System*, (Macmillan, London, 1973); *The Eurobond Market,* (Macmillan, London, 1973); *Parallel Money Markets — Volume One: The New Markets in London,* (Macmillan, London, 1972)

E. Chalmers (Editor), *Readings in the Eurodollar Market,* (Griffith & Sons, London, 1969)

E. W. Clendenning, *The Eurodollar Market,* (Oxford University Press, 1970)

Index

Note: *q.v.* in a sub-entry indicates that the subject is covered in an entry in its own right

Index

York, euromarkets, LIBOR, etc, and
individual countries
Finland 94-95
fire insurance 167, 178 .
fiscal policy, see taxation and individual
countries
fixed rate loans 234
floating rate loans 230-231, 238, 244
floor price 27
foreclosure (loan default) 84, 123,
233-234
foreign bonds 245
Foreign Credit Insurance Association
(FCIA) 198
foreign exchange gains and losses,
taxation of 158
foreign exchange risks 102-103, 193,
194
Forties field 65, 81, 240-241
France 4; export credits 183, 188, 191,
196, 197, 199, 200,201; financial
markets 207; government assisted ship
finance 94-95; marine pollution
174-175; taxation 233
Frankel, P. H. 28
Frankfurt Stock Exchange 207
freight insurance 105, 175-176
fuel oil 9-10, 13
Gabon, see also OPEC 4, 7, 11, 14, 17, 19
gas 5, 8, 125, 126; associated gas 125;
LNG 126, 136, 137; natural gas 7, 8,
125, 126; non-associated gas 125
gas injection 83
gas oil 9
gasoline 3, 9, 10
gathering station 6
Geneva Convention on the Continental
Shelf 148
geologists 56
Germany 4; export credits 188, 194-196,
199, 200, 201; financial markets 207;
government assisted ship finance .
94-95; marine pollution 175; taxation
147-148
s.s. Gluckauf 91
government involvement in oil industry
17-18, 69-71, 89; blocked cash 48;
fifty/fifty profit sharing 19; individual
countries, q.v.; participation 17, 21, 44,
67, 71, 89; revenues 20-21, 69-70;
state oil companies 17-18, 37, 67
government sponsored financial
assistance 181-182; for exports 183-199;
for North Sea 181, 197; for tankers
95-96
grace period (of bonds) 243
Greek fire 2
Greek tanker owners 93, 98
growth of petroleum industry 3
guarantees 104, 113, 116, 123; by banks

82-83, 104, 231-232; by government 82,
104; by holding company 231, 240-241;
completion guarantee 35, 65, 118;
performance bond 66-68, 80; project
finance, q.v.; take or pay agreement
120
Gulbenkian 28
Gulf Oil Corporation 16
Hamilton Bros 56
Hermes Versicherungs A.G. ('Hermes')
194
Hewitt field 213
historical development 2-3, 74-75, 91-92,
110-112
Holland, see Netherlands
Hong Kong tanker owners 98
hot oil 75
hull insurance 104-105
hydro-electricity 8
illuminating oils 2, 3, 7
illustrative agreement 61
importing countries 4
imputation system 148
independent oil companies 39;
exploration financing 56-63;
illustrative agreement 61; individual
companies, q.v.; ownership 16-18;
production financing 62-69
independent tanker owners: country of
registration 98; financing
arrangements 93-96; individual countries, q.v.;
marine pollution liability 174-175;
one ship company 98
India 4, 9
Indonesia, see also OPEC, 4, 11, 13-14,
17, 19, 23
insurance: by integrated oil companies
41-42; export credits 184, 185, 193;
purpose 139, 162
insurance business: claims adjusters 163,
165; classification societies 163, 166,
171; commercial insurance 162;
deductibles 164, 167, 171, 174;
insurance brokers 164; Lloyds 162-164;
mutual association 162-163; self
insurance 41, 163; surveyors 163-165,
178; underwriters 163
insurance cover (by kind of risk):
business interruption 167, 171;
commercial risks 184, 193-199; cost
escalation 194; earthquake,
windstorm and flood 168; export
finance 200, 201; fire, lightning and
explosion 167, 178; foreign exchange
193, 194; freight 105, 175-176; general
liabilities 166; loss of profits 167;
marine pollution 174-175; political
risks 184, 185, 193-199; products
liability 167; protection and indemnity
(P&I) 105, 171-172, 176; war risks 105

250